# Environmental Noise Barriers

2nd edition

# Environmental Noise Barriers

A guide to their acoustic and
visual design

Second edition

**Benz Kotzen and Colin English**

Routledge
Taylor & Francis Group

LONDON AND NEW YORK

First published 1999 by E & FN Spon

2 Park Square, Milton Park, Abingdon, Oxfordshire OX14 4RN
52 Vanderbilt Avenue, New York, NY 10017

*Routledge is an imprint of the Taylor & Francis Group, an informa business*

First issued in paperback 2019

*British Library Cataloguing in Publication Data*
A catalogue record for this book is available from the British Library

*Library of Congress Cataloging-in-Publication Data*
Kotzen, Benz.
   Environmental noise barriers: a guide to their acoustic and visual design / Benz Kotzen and Colin English. — 2nd ed.
      p. cm.
   Includes bibliographical references and index.
1. Noise barriers—Design and construction. 2. Traffic noise. I. English, C. E. (Colin E.) II. Title.
   TD893.6.T7K585 2009
   625.7'9—dc22                                              2008032407


ISBN13: 978-0-415-43708-0 (hbk)
ISBN13: 978-0-367-86522-1 (pbk)

# Contents

# Preface to second edition

Undertaking the research for this second edition has allowed us to revisit many of those people whose help and advice proved so valuable when writing the first edition. Travelling to see new barriers inevitably took us past old friends and allowed us to see how well these structures had stood the test of time. Once again we were struck by the wide discrepancy of provision from country to country. European countries such as the Netherlands, Switzerland, Austria and Italy continue to build substantial barriers to enhance the living conditions of their populations living near major roads, while, further afield, Australia is also creating some impressive barriers. Curiously, it is not always the poorer countries that fail to invest in good barriers to achieve acceptable environments, and for some unknown reason some countries simply lack the political will to invest in noise mitigation on behalf of their people.

Our travels allowed us to view many of those barriers that had made the greatest impression on us a decade ago. These continue to work well acoustically, but we were surprised to see how dated the more elaborate designs looked after ten years' service. It was often the simpler designs which appeared to fit into the current landscape better. Sadly, the ravages of time and graffitists have all too often taken their toll and this reinforced our opinion that, as with any built structure, barriers need occasional cleaning and maintenance. Planting is still considered important in barrier integration and is still one of the best ways of deterring graffiti.

As before, it is the Netherlands that leads the way in barrier development and innovation. For many years air quality and noise have both been seen as environmental impacts of major roads and it was exciting to see that important work is being undertaken in the Netherlands to design noise barriers that can reduce the air pollution caused by the road traffic. In the past a few barriers have been designed with additions to exploit solar energy. This tentative exploration for amalgamating barrier design with photovoltaics has seen a sea change in the design of the 1,200 metre long 620 kilowatt photovoltaic barrier at Freising, Germany. It is conceivable that many barriers designed in the near future will have a number of design functions apart from mitigating noise.

Perhaps the most exciting development that we encountered was the integration of noise barriers with non noise-sensitive buildings. All communities need a range of shops, sports facilities and workplaces that can be used as a buffer zone between the road and the residential area. It is puzzling that it has taken so long for the idea to become a reality, where economic as well as environmental positives may be achieved. This is particularly evident in the development of two striking barriers which incorporate showrooms, shops and restaurants in the Leidsche Rijn area of Holland. Contemporary examples we have investigated prove the case for treating environmental noise barriers as architecture and not street furniture. Additionally, respect for the genius loci (sense of place and context) is important, and barriers that are designed to fit in or, where possible, designed to make a visual statement appear as positive contributions in the landscape.

While progress in the UK has undoubtedly been slow, it is noteworthy that a programme for quietening existing noisy roads has now been introduced. Until this development it was only new and altered roads that were considered for noise mitigation. This change in policy has to be welcome, but we question the way in which the modest budget is being spent. The criteria for site selection were developed without any public consultation and the process has resulted in the installation of barriers of modest proportions in some of the noisiest location in the country. All too often these barriers appear to provide no meaningful improvements in the living conditions of those who they are meant to help and we are left with the inescapable conclusion that the process is skewed to allow politicians to claim that they have benefited the maximum number of people. Surely it would have been better to make substantial improvements in the conditions for fewer people?

Although the major advances in barrier design continue to be made overseas, the provision of barriers in the UK has also advanced. Taller barriers are being introduced and the proportion of sound absorptive barriers being constructed has significantly increased. However, it is clear that the political will to reduce environmental noise is as elusive as before. In the last decade the word soundscape has been coined, reflecting the growing interest in and awareness of the aural environment. We remain optimistic that this growing interest, together with the pressure of European harmonisation of noise policy, will result in progressive improvements in the design and provision of noise barriers in the UK.

The publication of the second edition of this book ten years after the first edition comes at a time where there are great changes in thinking on barriers, particularly where barriers can be used for air pollution control and where they can also provide energy for local communities. However, the idea that barriers can be fully incorporated as part and parcel of the economic generator for development is the most important advance in barrier thinking. Where new developments are being designed, environmental noise barriers no longer need to be considered a financial negative. The opposite is true.

The barrier can be seen as a starting point for innovative solutions with the provision of shops, showrooms and community facilities, which are expediently and strategically well placed alongside existing and future transport corridors.

Benz Kotzen and Colin English,
July 2008

# Preface to first edition

The genesis of this book was a jointly written conference paper titled 'Integrating European Scale Barriers into the English Landscape'. As we struggled to condense our material to fit a 40-minute slot, one of us rashly suggested that it might be easier to expand it into a book. Throughout this project we have been sustained by a desire to see an improvement in the current standard of practice in the control of road and rail noise in the UK and to ensure that those whose lives are afflicted by excessive traffic noise may enjoy a less stressful and more tranquil environment.

Work began on the project in the early 1990s. At that time the UK Government was pursuing a major roads programme which included the construction of many new bypasses and extensive motorway widening schemes. It became clear to us as we worked on some of these schemes, however, that the full noise implications were not being addressed. These roads were being designed to carry far greater volumes of traffic than had previously been the case, but the measures in use at that time were an inadequate solution to control the resultant noise. Taller and taller wooden fences did not offer a satisfactory solution to the noise problem; moreover, they would be an eyesore.

We had to look outside the UK to find more acceptable solutions to the problem. It soon became apparent that in certain European countries, although barriers were being used along roads to control traffic noise, these were quite different in specification to those in use in the UK. We were impressed, too, by the variety of barriers in use, many of them visually striking. We felt that it was important to research terrestrial transport noise control methods in Europe and elsewhere, and ask key questions. Could similar barriers be employed in the UK? Why is the use of sound absorptive barriers widespread abroad and seldom considered in the UK? Why are some barriers visually acceptable while others offend the eye?

All too often in the UK, the impression is given that barriers are not properly planned and designed. Good design should involve acoustic engineers and landscape architects who may have conflicting priorities. Accordingly, we have set out to examine both acoustic and landscape issues, to give guidance on good practice and to highlight the pitfalls of bad design.

The Department of Transport issues guidance for its own road designers on how to plan, design and build barriers and this can also be followed by those building barriers on local roads and railways. Our book is intended to supplement that advice. We want to show the variety of solutions which can be achieved in different locations. To do this we have included as many photographs and illustrations as possible, not so that these designs can simply be copied, but to serve as a resource and stimulus in future noise barrier design.

Environmental noise pollution is a problem that is being addressed with increasing seriousness in many countries in continental Europe. The UK has much to learn from these countries about noise control policies and their implementation, since their solutions offer variety and ingenuity of design. Our studies have taken us to Belgium, Denmark, France, Germany, Holland, Italy and Switzerland in order to examine the wealth of barriers devised for different acoustic and landscape situations in these countries. We have, of course, met many transport industry officials and planners during our travels and been impressed by their sense of social responsibility and their commitment to the task of providing a better acoustic and visual environment for their citizens. Moreover, during meetings with continental barrier manufacturers, we have been encouraged to learn that most have close links with manufacturers and suppliers in this country who offer the same products and expertise. This will allow similar solutions to be readily implemented here.

Sadly, short-term solutions and cost-cutting tend to dog road planning in the UK. A recently constructed road scheme has cost the Government more in compensation payments to local residents for the depreciation caused to their property than the original cost of building the road. It is true that since we started work on this book the scale of the roads programme has been reduced considerably; the need for larger, better-designed barriers has not diminished, however. The public, increasingly well informed about the problem of excessive noise, is no longer prepared to tolerate a noisy environment. Compensation costs will continue to grow unless and until the impact of traffic noise is reduced.

Furthermore, the reduction of noise on existing roads is long overdue. The very high noise levels endured by so many people living alongside established roads is becoming intolerable. Many of these roads now carry far more traffic than was ever anticipated. We should follow the example of many other European countries by recognising the damaging social and health effects caused by such exposure and implementing noise control programmes on our busier roads.

Our original purpose in writing this book was to create a source book for all those involved in the design of barriers in the UK. We felt, too, that they would be interested in our investigation of the background reasons for the slower development of UK barrier design compared with elsewhere. Nevertheless, we hope that, since the design principles in this book have universal application, it will be of use to barrier designers outside the UK too.

It is possible to control terrestrial transportation noise, and to improve the appearance of the necessary barriers. There are many examples of this in Europe and elsewhere; more recently, there have been some encouraging developments in the design and provision in the UK. We hope that this book helps to stimulate debate on the better protection of those affected by traffic noise and acts as a catalyst for the design and implementation of successful barriers.

Colin English and Benz Kotzen,
July 1998.

# Acknowledgements

As with the first edition of this publication, we are grateful to the many people, companies and organisations that have willingly shared their time, knowledge and expertise in researching the second edition, making it vital and contemporary. Many of those who assisted our pursuit of knowledge when writing the first edition have been as generous the second time round. Particular mention should be made of Cristo Padmos of the Rijkswaterstaat, Delft, the Netherlands, who munificently provided time for meetings, disseminating information, field visits and a transfer of knowledge on numerous occasions. Thanks are also due to many individuals and their companies who have met us and shared information on their own products, but have also generously shown us other interesting examples. This open-minded approach has assisted the writing of this book, where good ideas and methods are celebrated.

We would thus particularly like to thank Joop Van Campen of Van Campen Aluminium, Lelystad, the Netherlands; Huub Maas of Evonik Degussa International AG, the Netherlands; Gordon Smith of Clarke and Spears International, Woking, UK; David Fresco at Kokosystems, Poeldijk, the Netherlands; Pamela Lowery, Highways Agency, UK; Lotje van Ooststroom, Rijkswaterstaat, the Netherlands; and David van Zelm van Eldik, Ministeries van V&W, Vrom en LNV, the Netherlands.

Most of the photographs have been taken by Benz Kotzen and Colin English. In some cases, however, this has not been possible and a few photographs and illustrations have been provided by other people, companies and organisations. In this regard we thank Barbara Vanhooreweder of the Vlaamse Overheid; Jean-Luc Le Gouallec for Patrick Blanc; Gregg Watts; Karel Novotny, Novotech, Czech Republic; Prof. ir. Fons Vewrheijen, VVKH Architecten, Leiden, the Netherlands; Ateliers Jean Nouvel, Roma, Italy; Ole Refshauge, Milewide A/S, Harderslev, Denmark; Tonomao Okubo at the Kobayasi Institute of Physical Research; and Tonkin Zulaikha Greer Architects, Surry Hills, NSW, Australia.

# Location of photographs

# Introduction

## Background

The growth in the use of noise barriers across Europe, the USA, Australia and the Far East, reflects the growing concern of the general public about noise pollution caused by major infrastructure projects, in particular roads and railways. This concern about the adverse effects of noise in the environment has, in turn, led governments to create the legislative framework that has motivated the responsible authorities to mitigate noise in urban, semi-rural and even rural localities.

The growing demand for a quieter environment has caused the noise barrier market to grow considerably in recent years across Europe and around the world. This, no doubt, will continue, as project designers respond to increasing environmental concerns and have to comply with improved noise attenuation guidelines, and more demanding legislation. The acoustic and aesthetic standards of barriers have improved considerably and will continue to improve at an accelerated rate as information and expertise is disseminated across borders. These pressures have been given added impetus with the publication of European design standards and implementation of the European Union's programme for harmonisation of noise standards across Europe.

It is often tempting to regard traffic noise as a 20th-century phenomenon, but nothing could be further from the truth. The Romans were all too familiar with the unwanted noise of wheels on stone streets and issued a decree which banned the use of chariots on the streets of Rome at night. Sadly for the Romans, their leaders did not have a monopoly of acoustic wisdom: Julius Caesar passed a law which required all goods deliveries in Rome to be made at night. Not content with creating this noise nuisance, Claudius subsequently extended the law to all towns in Italy and Marcus Aurelias imposed it on every town in the Roman Empire.

The fall of the Roman Empire heralded an apparent decline in interest in controlling the adverse effects of traffic noise. With the growth of cities in Tudor Britain vehicle noise again became a problem. In 1586 an Act was passed in London whereby fines were imposed for any coach or cart heard to creak or 'pype' through want of oil.[1] There then appears to be little reference to the problem for several centuries until the invention of the internal combustion engine irrevocably changed the aural landscape. It took only three decades of motor transport to convince the British Government to introduce legislation to control the noise emitted by motor vehicles. A simple act, passed in 1929, predated the invention of any means of measuring noise: it relied on a policeman's and then the court's judgement to decide if the offending vehicle was too noisy.[2] Despite their capacity to create high noise levels, railways remained immune from noise legislation until 1996, when the resurgence of railway building necessitated the introduction of regulations requiring noise insulation to be provided for houses badly affected by noise from new railways. This in turn led to the provision of noise barriers along some sections of new railways.

Traffic noise barriers have been installed in the UK for at least the last 40 years. Initially, barriers comprised walls constructed from horizontally stacked concrete planks and often these were faced with intermittently spaced vertical timber slats, although these appear to have served little acoustic or aesthetic function. More typically, major road schemes across the country have utilised earth mounds and timber fences to mitigate noise and visual intrusion: in certain locations, however, these natural-looking barriers are visually out of place. There continues to be a need to design such barriers, now termed 'environmental barriers', in appropriate, forward-looking and environmentally conscious ways.

There remains a particular need for this change in attitude to design in the UK, where noise barrier design and provision still lags behind both the achievements and plans in many other countries. There are many different reasons why the development of noise barriers in the UK has been less ambitious and extensive than in other European countries. Four of the main reasons are:

- the unforeseen effects of the Land Compensation Act, 1973;[3]
- the different targets for community noise levels set by each country;
- the official traffic noise calculation procedure tends to overestimate barrier performance;
- the lack of a programme to reduce noise on existing roads in the UK prior to 2000.

The Land Compensation Act was introduced to provide compensation to owners of property which was devalued as a consequence of public works, irrespective of whether the property or part of the property was needed for the execution of those works. Under the Act, regulations were introduced in 1975 which enabled part of the compensation for householders to be the provision of noise insulation.[4] This is available where the exposure to noise is increased to 68 $dBL_{A10,18hr}$ due to a new or altered road, and where certain other criteria are met.

Throughout the time that the Land Compensation Act has been on the statute book, it has been the policy of the UK Government to provide screening in preference to noise insulation. All too often the cost of barriers was balanced solely against the cost of noise insulation and this simple fiscal test resulted in insulation being offered to owners of affected properties. It is worth noting, however, that a preference for screening was reaffirmed in 1994 by the Government in its planning guidance on noise given in PPG 24 which remains extant.[5]

The Land Compensation Act was an enlightened piece of legislation and there is much to commend in it. However, it is an unfortunate consequence of the subsequent Noise Insulation Regulations that the qualification threshold noise level has become a *de facto* design target level in the absence of any other official guidance. In contrast, many other European countries have introduced legislation on design targets which are often significantly lower than the 68 dB $L_{A10,18hr}$ level used in the UK. Notable among these are the Netherlands and Denmark, where the permitted level is 55 dB $L_{Aeq,12hr}$, a target which results in environmental noise which is, subjectively, only half as loud as the level permitted in the UK. Where target levels are lower, it follows that screening will be more robust and substantial. It also means that in those countries the use of screening is more widespread and that greater experience has been gained in the design, manufacture and construction of noise barriers.

In the UK, calculations of road traffic noise levels are made following the Department of Transport's method.[6] This models traffic as a single line source for all roads, except those dual carriageways where the central reserve is unusually wide or the carriageways are at different heights. The single noise source is located 3.5 m in from the nearside edge of the carriageway, which is a fair approximation for single and narrower dual carriageway roads. Where barriers are used, this model effectively places all of the traffic close to the barrier, where it will receive the maximum benefit from the barrier. Studies have shown that this can lead to an overestimate of barrier performance of over 2 dB for wide dual carriageway roads.[7]

Another reason for the absence of substantial barriers in the UK is that prior to 2000 there was no policy to reduce noise exposure from existing roads. Such policies, often known as 'black spot' policies, are common in the rest of Europe; many countries often spend as much on quietening existing roads as on the control of noise from new schemes. Historically, the official reason for not adopting a black spot policy in the UK has been that owners of properties along these noisy roads would have been compensated at the time of the construction and, therefore, it would be offering double compensation to also provide screening. However, this policy did nothing for those people living on roads constructed before the introduction of the Land Compensation Act, nor did it help those people affected by noise from roads where the traffic growth has been much greater than forecast. In 1998 the Government published a white paper in which it signalled its intention to create a ring-fenced annual budget for screening some of the noisiest existing roads.[8] Unfortunately, no noise reduction target was specified and this has meant that the small annual budget has been spent in providing barriers of modest scale, despite their use at sites described as the 'most serious and pressing cases'. These barriers have often been located at sites with housing on both sides of the road and the policy has given rise to a significant increase in the use of sound-absorptive barriers.

New major roads are generally planned to avoid residential areas and, therefore, the communities exposed to the highest levels of traffic noise are often found along established, heavily trafficked roads within cities and other developed areas. In these situations housing is often very close to the road and reducing noise would require the use of tall barriers. In such locations, timber fencing, which is the most frequently used system in the UK, is often visually inappropriate. This problem has been recognised by the authorities in those countries which try to control noise in these areas, and thus a range of more suitable materials and barrier forms has been explored and developed.

The long-standing reluctance to tackle the relentless increase in traffic noise in the UK should have had some benefit. During this period a vast range of barrier types has been used and tested abroad. The authorities and designers in the UK now know what can be achieved, in terms of acoustic mitigation, visual aesthetics of the barriers, and the effect these, often large-scale, constructions can have in different environments. However, this knowledge seldom leads to innovative design and timber remains the material generally used for new barriers. Although these timber barriers have increased in scale, length and occurrence in the landscape and many utilise sound absorptive panels, they are generally not well integrated into the environment. After 40 years, barrier provision in the UK still lags far behind that of many other countries. The issue of traffic noise is not high on the Government's agenda and thus traffic noise remains a particular problem, especially on urban routes. The predominant provision of timber barriers, usually without planting, illustrates that the barriers are not designed to fit in with a particular genius loci (sense of place.) Although it may have been expected that there would have been a real public demand for the provision of well-designed and appropriate structures in the landscape in the UK, this has unfortunately not materialised.

## Environmental noise barrier terminology

One of the primary objectives of this publication is to encourage a universal and appropriate use of terminology for describing noise barriers which are variously called noise fences and screens, or acoustic barriers and, more recently, environmental barriers.

The lack of an agreed terminology led to considerable deliberation over the title for this book as the current use of terminology in this field can often be misleading. For example, the term 'environmental barrier', signifies a noise or visual barrier to the Highways Agency and its agents in the UK, but in some countries it may well mean some kind of geotextile barrier which isolates contaminated soil. Other professions may also use the term differently, ecologists it may use it to describe fencing that could protect wildlife such as deer, badgers or newts from entering a road corridor.

It also appears that the terms used to describe barriers could disguise their purpose, thus confusing designers and clients alike and this makes it more difficult for the public to understand the objectives and intentions of the proposals. This situation helps to create a climate of misunderstanding and a lack of clarity of purpose. At worst, it causes lack of trust between the developer and the public. It is, therefore, extremely important to categorise these barriers/screens/fences according to their function to make sure that everyone understands their purpose.

Generally, the Highways Agency categorises noise barriers and landscape or visual barriers by the generic term 'environmental barriers'. This term encompasses all barrier/fence/screen/mound types, regardless of whether their function is to mitigate noise or to protect views and/or the landscape. It should be stressed that environmental barriers in the UK are thus designed to reduce noise and to protect views in the landscape.

To avoid further confusion, it is also important at the outset to use the appropriate 'aural' terminology in describing a barrier, for they are often called acoustic barriers or sound walls as well as noise barriers. Since it is common to define noise as unwanted sound, it seems appropriate, therefore, to use the term 'noise barriers' as their function is to attenuate a specific traffic noise problem. Thus, the general term 'environmental noise barrier' may be used to encompass every type of structure used to reduce noise.

Alternatively, where a barrier is designed to mitigate certain unwanted views, or to screen a disturbing alteration or addition to the landscape, it may best be termed a 'visual barrier'. It is often the case that the barrier is intended to mitigate both noise and views, since where there is a noise problem there is usually a visual problem as well. In this case it should be termed a 'noise/visual barrier'. This, however, is not as simple as it may seem, for it is also important to recognise that, although a barrier may have a dual function, very often the heights necessary for each function and the type of barrier required may need to be different.

A certainty of function is especially important when considering materials, since these may have an effect on the height of the barrier. Due consideration must be given to its horizontal alignment and its appearance, because the barrier will be subject to close scrutiny throughout the design and planning processes. Without this clarity of function, the issues are clouded and this may result in the design of an inappropriate barrier.

There is also some confusion as to whether a barrier should be called a fence or a screen or perhaps just a barrier, as is evidenced by the current use of the terms noise fence, noise screen and noise barrier. Although, in many cases, it may be possible to term a barrier a screen or a fence, in some cases it is inappropriate since it suggests a much less substantial structure than would be needed to control noise adequately.

The above discussion does not take account of terminologies used to describe noise barriers in other languages and countries and the discussion may thus not be relevant in some parts of the world. In the USA they are known as 'sound walls'. The European Environment Agency provides a useful 'Environmental Terminology Service',[9] which translates the English term 'noise barrier' as noted in Table 1.1.

**Table 1.1** Translation of the term 'noise barrier' into 22 European languages

| | |
|---|---|
| Bulgarian: | преграда за противошумова защита |
| Czech: | hluková bariéra |
| Danish: | støjvold |
| Dutch: | geluidsafscherming, geluidswal |
| Estonian: | müratõke |
| Finnish: | melueste |
| French: | écran antibruit |
| German: | Lärmschutzeinrichtung, Lärmschutzwand |
| Greek: | ηχοπέτασμα |
| Hungarian: | {NOI} zajvédõ (fal), zajárnyékoló fal |
| Icelandic: | hljóðmön |
| Italian: | barriera antirumore |
| Latvian: | trokšņa slāpētājbarjera, prettrokšņa barjera |
| Lithuanian: | triukšmo barjeras |
| Norwegian: | støyskjerm |
| Polish: | ekran przeciwhałasowy |
| Portuguese: | barreira acústica |
| Romanian: | Barieră sonoră |
| Slovak: | hluková bariéra |
| Slovenian: | pregrade za zaščito pred hrupom |
| Spanish: | barreras acústicas |
| Swedish: | bullerskärm |

## Noise is a landscape issue

Although the emphasis of this book is on noise and noise barriers, noise is also a landscape issue,[10] in that it has a great impact on the perception of the character and quality of the landscape. A landscape assessment may describe the character of a landscape as having a quiet or tranquil setting, or as being noisy. This affects the categorisation of the landscape quality of an area, which may deteriorate with the presence of noise. Whereas some people may concentrate on the negative impact that barriers have on the landscape due to their scale, appearance or other perceived undesirable qualities, the noise itself may well have an adverse effect on people's enjoyment of the landscape and can have, therefore, an adverse affect on landscape quality, landscape character and the quality of life. The experience of sitting in a garden listening to bird song is quite different from and more pleasant than sitting in a garden which is dominated by the roar of traffic noise. In rural environments too, enjoyment of the landscape and leisure activities may be diminished by the presence of noise. More often than not, noise is the first and only indicator of the presence of development or infrastructure projects, as these schemes are often well screened by planting.

Although the reduction of noise in a given location through the use of barriers could help to improve the problem which a development has caused to the environment, it may create others. It is important to acknowledge the effect these often large and imposing barriers may have on other environmental issues. They may affect views, light, microclimate, access, wildlife and birds. These structures, which may be 5–10 m high, or even in exceptional circumstances up to 20 m, should be integrated, as far as possible, into the local surroundings and all environmental issues relating to them be properly examined (Figures 1.1 and 1.2). A barrier should reduce noise to the required levels, and be acceptable to the planning authorities, but, to be truly

**1.1** Twenty metre high barrier protecting residential blocks on the Périphérique, Paris

**1.2** Twenty metre tall barrier protecting residents at the A23, Vienna

successful, it must merit approval from local inhabitants. In order to satisfy these conditions, the barrier must be designed to integrate well into its surroundings. When the wrong type of barrier is constructed, which degrades landscape character and diminishes landscape quality, it will inspire local animosity. Social surveys have shown that where this is allowed to happen the public's perception of any acoustic benefit is noticeably reduced. An extreme example of the problems caused by failing to consult local people occurred in Oregon where the highways authority was forced to remove a noise barrier because of local hostility.[11]

Landscape architects prefer not to see noise or visual barriers, as they make a significant impact on the landscape. Where the need for barriers outweighs the sum of the other negative environmental effects, it is the task of the landscape architect to help to improve the visual environment for those living adjacent to intrusive developments, and to maintain the integrity of the landscape. Aesthetic and visual integration is not a simple task, not least because in many cases there is likely to be opposition to barriers due to their scale and appearance. However, it is possible to satisfactorily integrate these structures into the landscape with sensitive design solutions: employing appropriate vertical and horizontal alignments; with sympathetic use of materials and combinations of materials; with creative and appropriate use of pattern, colour and texture; and especially with imaginative planting. When designed well, environmental noise barriers and visual barriers may be satisfactorily integrated into the landscape and be acceptable to local people.

It must also be remembered that environmental concerns are growing and that people no longer accept mediocre standards and poor-quality service. The importance of protecting the integrity of the countryside and people's enjoyment of it should not be underestimated. If new infrastructure is required, the necessary measures must be taken to lessen or negate potential negative impacts and this should be undertaken as an integral part of the works. Mitigation must be seen as an essential part of any scheme. In terms of landscape architecture, it forms a major part of the landscape objectives, strategy and design.

Barrier design is a complicated process. The best results are likely to be achieved through the co-ordinated services of qualified acousticians, civil and structural engineers, landscape architects and architects. Other professional expertise may also be required, including advice from geotechnical, ecological, irrigation, horticultural and other environmental and planning specialists.

It appears that the environmental thinking and practice relating to environmental noise control is less advanced in the UK than in many other parts of Europe. There are, however, signs that the British public is no longer prepared to accept the provision of inadequate environmental solutions. Continental European scale barriers have been installed at a few sites in the UK and a wider range of barrier materials can now be seen alongside roads, particularly on the busier motorways in southern England. The process of improvement of standards and provision is likely to receive a boost through the European harmonisation programme. The preparation of noise maps for major towns and transportation networks is now largely completed and the preparation of noise action plans has begun. These action plans are to be written by each member state, but if states conform with each other, this process may mark a welcome shift in emphasis away from compensation in favour of better provision of noise mitigation.

## Barrier use: a contrast in provision

It makes good sense to include local residents and authorities in the planning of barriers at the earliest stage possible, as is the case in the Netherlands or Denmark, for example. It is beneficial in terms of cost, the smooth running of the project and to dispel misunderstanding to work within a framework of trust and consultation. At present in the UK the public is included in this process at too late a stage and consequently feels marginalised and helpless. A sea-change is needed to ensure that local people are involved in this process early enough. Of course, the design process should not be annexed by the public, nor should barriers be designed by a committee of laymen; nevertheless, people affected by new or extended traffic systems should feel that they have a say in the solutions and should be kept fully informed about developments.

When considering the problems of noise and visual intrusion, it is also important to look at alternative methods of mitigation other than through the use of barriers, including the use of quiet road surfaces, the insulation of properties or even tunnelling. For the reduction of noise, all these options need to be considered individually and in conjunction with one another, as the optimum noise and landscape mitigation strategy may involve using a number of solutions (Figure 1.3). In the Netherlands, for example, the extensive use of porous asphalt has significantly reduced road traffic noise (Figures 1.4 and 1.5). Reducing tyre/surface noise in this way allows designers and planners to keep barriers heights to a minimum. This, in turn, may minimise environmental effects and reduce the overall capital costs of a scheme. At present, over 70 per cent of Dutch motorways are surfaced in porous asphalt. It is the intention that, by 2012, the whole network will be treated similarly, as a response to an increase in the maximum speeds on motorways in Holland from 100 kph to 120 kph which increased noise

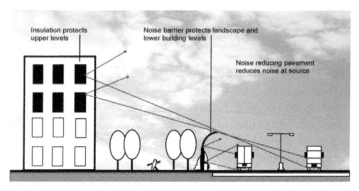

**1.3** Composite noise reduction

**1.4** Core samples of running surfaces. Left to right: twin layers of porous asphalt with rubber coating, single layer of porous asphalt with rubber coating, single layer porous asphalt, porous concrete, dense asphalt

**1.5** Edge of porous asphalt running surface

levels by some 2 dB. The Dutch Noise Abatement Act states that 'measures should be taken to level down the increase of the noise level (stand still principle)'.[12] In the UK porous asphalt gained some popularity in the 1990s, but durability problems had lead to a switch to the use of thin wearing courses such as stone mastic asphalt, which are 2–3 dB quieter than hot rolled asphalt. The Japanese have developed a 'Porous Elastic Road Surface' (PERS / 媒). The PERS is composed of treated silica sand and granulated rubber made from old tyres as the aggregate and urethane resin as its binder.[13]

Compared with the UK, noise barriers in countries such as Germany, the Netherlands and France form a greater part of the urban and semirural landscape and have done so for several decades. Absorptive barriers, namely barriers which comprise materials which absorb noise, have been commonly used throughout much of Europe in and around cities for many years (Figure 1.6), whereas in the UK the construction of low, timber, sound-reflective fences has been the norm. In much of continental Europe care has been taken not only to provide barriers of the correct scale but also to investigate the use of a wide range of materials and construction techniques. Of course, mistakes have been made and now many barriers appear dated and strangely out of place in the landscape. However, credit should be given to the planners responsible for their commitment to protecting people and the environment from unwanted noise and disturbing views.

**1.6** Mature aluminium absorptive barrier, Hamburg

Since the early days of noise barrier installation, great steps forward have also been taken in continental Europe with regard to noise reduction target levels, noise barrier design and appropriate consultation processes. In the late 1990s, Denmark provided a fine example of good design practice when the growth of traffic on the elevated motorway through the Bispeengbuen area of Copenhagen was found to have become excessive. In 1995 a comprehensive social survey of residents was carried out and then a select group of four design companies was invited to submit schemes to provide a minimum of 8 dB reduction in noise at the nearest dwellings. Sections of each of the proposed barriers were built *in situ* (Figures 1.7–1.10). The public was then

**1.7** Competition entry 1996, Bispeengbuen, Copenhagen

**1.8** Competition entry 1996, Bispeengbuen, Copenhagen

**1.9** Competition entry 1996, Bispeengbuen, Copenhagen

**1.10** Competition entry 1996, Bispeengbuen, Copenhagen

asked to comment on the design and an expert panel made the final selection. It is worth noting that a parallel competition was held to design street architecture and to enhance the austere area beneath the motorway. The winning design, a simple, but elegant, transparent screen with an extruded aluminium support structure, was built in 1998 (Figure 1.11).

Although environmental noise barrier construction could be regarded as a necessary evil, in continental Europe every effort is made to make a virtue out of necessity. Barriers are now part of the fabric of the environment, so planners, in consultation with environmental agencies and local residents, take great care to make them blend with their surroundings as well as ensuring that they provide effective noise mitigation. Although some barrier solutions are far from perfect and their construction raises questions about severance, personal safety, and light, particularly in urban areas, the fact that they have been erected shows a practical commitment to public well-being and the integrity of the environment. The case for environmental responsibility is gaining public

**1.11** Winning design installed at Bispeengbuen, Copenhagen

**1.12** Prestigious car dealership commercially strategically located alongside a Dutch motorway forms part of the noise control for the new Leidsche Rijn development at Utrecht

support; environmental noise barriers should gain acceptance as people learn more about the issues at stake. Barriers have not generally been regarded as prestigious structures, but this does not mean that they should not have been designed with care. The fact that they will be seen by millions of people every year is realised by forward-thinking planners and developers. Not only can they have a significant effect on the public's aural and aesthetic appreciation of the landscape beyond the barrier but also they provide opportunities for development and commercial activity (Figure 1.12).

The landscape of the British Isles is like no other, particularly in terms of its intimate scale and varied character, so what is built elsewhere cannot simply be translocated. It has often been argued by planners in the UK that the scale of continental European barriers is often inappropriate to the smaller-scale landscapes of the British Isles. The argument that continental European solutions and designs cannot be adapted is now largely discredited, based as it is on a lack of imagination, laziness and a reluctance to accept the scale of the problems presented by today's transport corridors. It is also an excuse for failing to allocate the appropriate budget for environmental protection. The difference in standards in the UK compared with elsewhere in Europe may suggest that the British people are less disturbed by noise pollution. No studies have revealed any differences in the effects of noise on the daily quality of life or health of all individuals of different nationalities, however. Indeed, the United Nations' World Health Organisation has issued its own guidelines on acceptable environmental noise which are uniform throughout the world.[14] The OECD reported that traffic noise at typical levels of emission does not cause any immediate risk of hearing loss, but there are other important nonauditory negative effects.[15]

If the environment is to be improved for those people who are affected by transport noise, it is necessary to see and understand what can be achieved by studying the potentials of materials, contemporary and past examples and the commercial possibilities of what has been realised elsewhere. Furthermore, continuing changes in attitude and legislation will be needed to ensure that a quieter, yet visually acceptable environment is provided both in cities and in the countryside.

## References

1. Bennett. E. (1982) *The Worshipful Company of Carmen of London*, Barracuda for the Company, London, p. 26.
2. *The Motor Cars (Excessive Noise) Regulations, 1929, No. 416*, HMSO, London.
3. *Land Compensation Act* (1973) HMSO, London.
4. Department of the Environment (1975) *Statutory Instruments, 1975, No.1763 Buildings. The Noise Insulation Regulations*, HMSO, London.
5. Communities and Local Government (1994) *PPG24: Planning and Noise*, HMSO, London.
6. Department of Transport, Welsh Office (1988) *Calculation of Road Traffic Noise*, HMSO, London.
7. English, C. and Swift, C. (1993) *Assessing Noise of Wide Motorways*, Proceedings of the Institute of Acoustics 15(4), pp. 807–14.
8. Department of Environment, Transport and the Regions (1998) *A New Deal for Trunk Roads in England*, HMSO, London.
9. European Environment Agency Terminology Service – http://glossary.eea.europa .eu/EEAGlossary/N/noise_barrier.
10. English, C. and Kotzen, B. (1994) *Integrating European Scale Barriers into the English Landscape*, Proceedings of a joint IHT/ENBA seminar: *Environmental Noise Barriers – A New Perspective*, 10 November 1994, Maidenhead, England.
11. The Transportation Research Board (1982) *Highway Noise Barriers*, National Academy of Sciences, Washington, DC, p. 11.
12. Padmos, C.J. (1995) *Development of Low Noise Surfacing in the Netherlands*, International Conference on Roadside Noise Abatement, November 1995, Madrid.
13. Information and photograph from the Yokohama Rubber Company, Ltd, Tokyo, Japan.
14. World Health Organisation (1999) *Guidelines for Community Noise*, WHO, Geneva.
15. Organisation for Economic Co-operation and Development (1995) *Roadside Noise Abatement*, OECD, Paris, p. 20.

# Defining the need for barriers

# 2

## Legislation and policy

### Introduction

The provision of a traffic noise barrier on any road depends on both the relevant legislation and on the adopted policies of the highways authority concerned. In general, the design and construction of all public roads in the UK must comply with the legislation extant at the time of granting of orders for its construction, but policy regarding the provision of traffic noise barriers varies according to which authority is responsible for the road.

The current legislation in the UK concentrates primarily on providing a framework for compensating those adversely affected by road traffic noise. Government policy, on the other hand, is increasingly addressing the need to reduce traffic noise at source.

Since the 1970s legislation and policy have concentrated on mitigating the effects of new schemes, and it is not surprising, therefore, that it has dealt exclusively with the effects of road traffic noise. The 1990s, due to a combination of environmental and economic pressures, saw a revival in the building of railways and tramways in the UK. Potentially, this could have led to inequitable situations, where those communities affected by new railway noise were disadvantaged when compared with those affected by new road noise. The Government's response was to introduce new regulations which were designed to treat those affected by train noise in an identical manner to those affected by road traffic noise. Initially, this approach seemed entirely reasonable, but implicit in it is the presumption that the prevailing policy with respect to road traffic noise was appropriate for the late 20th century and beyond. At a time when it was becoming increasingly clear that road traffic noise abatement in the UK was falling considerably short of the standards being achieved in many other countries, many saw this as a lost opportunity to improve the standards of provision.

### The Land Compensation Act, 1973

The Land Compensation Act, 1973[1] was introduced to provide a mechanism to compensate those who were adversely affected by execution or operation of public works. This, of course, included all road schemes and the Act has been pivotal in shaping the country's response to the problem of traffic noise. This Act made provision for homeowners to be offered noise insulation provided that certain qualification criteria were met. It also allowed for payments to be made where residential properties suffered a drop in value due to the adverse effects of a road scheme, including noise.

### The Noise Insulation Regulations, 1975

The Noise Insulation Regulations (amended 1988)[2,3] set out the qualification criteria which apply in England and Wales for determining those residential properties which may be offered noise insulation under the Land Compensation Act. Grants are only available for residential buildings within 300 m of the edge of the carriageway under construction or being altered. To qualify for a grant, properties affected by noise from a new road or carriageway must meet all three of the following tests:

- the total expected maximum traffic noise level within 15 years of opening of the road scheme, i.e. the relevant noise level, must not be less than the specified noise level (defined as 68 dB $L_{A10,18hr}$);
- the relevant noise level shall be at least 1 dB(A) more than the prevailing noise level, i.e. the total traffic noise level obtaining prior to the start of works to construct or improve the road;
- the contribution to the increase in the relevant noise level from the new or altered road must be at least 1 dB(A).

The regulations place a duty on the highway authority to offer grants, or to carry out noise insulation works, where the above conditions are met. However, separate provisions are made for properties which are affected by highway alterations which do not include the provision of new carriageways. In these cases, there is a discretionary power to provide noise insulation grants provided that, as a minimum requirement, the same three qualification tests are met. The discretionary nature of this power, with no clear guidance on when it should be exercised, inevitably leads to anomalies in its application. Many highways authorities adopt more stringent standards and require that there is a 3 dB increase in the relevant noise levels above the prevailing noise level, whereas others have required a 5 dB increase. However, for many motorway widening schemes, the Highways Agency has adopted the same criteria as used for new roads.

The regulations also enable noise insulation to be provided for dwellings which are seriously affected by road construction noise for a substantial period time. This again is a discretionary power and there are no published definitions of 'seriously affected' or 'substantial period', which can lead to discrepancies in application.

The equivalent regulations adopted in Scotland,[4] and in Northern Ireland,[5] differ from those described above in their treatment of the relevant noise level. Rather than forecasting the noisiest year within 15 years of opening, the highways authority is required to review the traffic noise from the road regularly and update the insulation accordingly. This has the advantage of subsequently insulating further properties affected by increased noise from roads which are subject to greater traffic flows than forecast prior to the opening of the road.

The Noise Insulation Regulations only provide powers to grant noise insulation for residential buildings, and while the Government indicated in 1990 that it was considering extending these powers to include hospitals and schools, it is a matter of some regret that no widening of the scope of the Regulations has been made.[6] Another aspect of the regulations which urgently requires modification is the 300 m limit for eligibility. When this was first introduced it reflected the fact that traffic conditions could not possibly cause the qualification threshold to be exceeded outside the area covered by this limit. It is notable that when the M1 was opened in 1959 it carried just 14,000 vehicles per day, whereas today the busiest sections of the motorway network carry in excess of 200,000 vehicles per day. Today's greater traffic flows and speeds now mean that properties considerably outside the 300 m distance limit could meet the noise criteria for insulation, but there is no mechanism which can be used to offer this to the affected properties.

**The Highways Act, 1980**

It has always been recognised that control of noise at source is more desirable than providing noise insulation at the affected properties. As traffic volumes and noise levels increase, the numbers of properties adversely affected grow and it becomes an increasingly viable option to provide screening for roads. Clause 246 of the Highways Act[7] gives highway authorities the powers to acquire land to mitigate the effects of a new or altered road. The purchase of such land must commence prior to the opening of the road scheme.

Clause 282 of the Act gives powers to a highways authority to provide or improve the traffic noise barriers on its own land at any time. There are no powers to acquire, by compulsory purchase, land for screening after the scheme is completed; however, this can be done with the agreement of the owner of the adjacent land.

**National Government policy**

The responsibility for the motorway and trunk road system rests with the Department for Transport (DfT, previously the Department of Transport, DoT) and, therefore, Government policy is directly applicable to these roads. In 1990 the Government summarised its policy on reducing road traffic noise in its White Paper, *This Common Inheritance*.[8] This stated that the lines and levels for roads were selected to minimise noise and that noise barriers and mounds were used to reduce noise. It also anticipated that quieter road surfaces would be used to reduce noise at source.

In 1992 the DoT permitted the use of porous asphalt in urban and noise sensitive locations. At the same time it banned the use of concrete running surfaces for roads carrying over 75,000 vehicles per day.[9] Earlier that year the DoT approved the use of exposed aggregate concrete, or whisper concrete.[10] The effect of these decisions is to allow the use of surfaces such as porous asphalt, which is 3–4 dB quieter than hot rolled asphalt and also whisper concrete, which is about 1 dB quieter than hot rolled asphalt, while banning the noisiest (concrete) surface. The use of porous asphalt was sometimes less than successful and alternative low-noise surfaces, such as stone mastic asphalt are now favoured for use in the UK.

The reductions in noise which can be provided by inherently quiet road surfaces are relatively modest when compared with what can be achieved with screening. As noted above, the increasing volumes of traffic and the concomitant increase in properties affected by noise has resulted in the more frequent use of noise barriers to control noise. There is no comprehensive policy for deciding when and where to use barriers. Instead, it is left to each project design team to decide where the use of barriers is justified. The DfT, through its Highways Agency, has issued guidance on the use of noise barriers in the *Design Manual for Roads and Bridges (DMRB) Volume 10*,[11] but, notwithstanding this, the *ad hoc* approach has led to a wide range of design standards being adopted. Often the objective is to ensure that noise levels are limited to just below the noise insulation qualification threshold; the results of this have often, and rightly, been described as maximising the noise nuisance. Some schemes use barriers to substantially improve the noise climate for nearby residents, whereas others use them merely to maintain noise levels at those obtaining prior to an improvement being implemented.

It is somewhat surprising that, among road designers, there remains a very cautious approach to the use of barriers, as the advice in *DMRB* is so very positive. In particular, the advice given on road improvement within the land-take is very strong and among the environment objectives it lists:[12]

- Minimise impact on natural and built environment;
- Minimise noise disturbance.

When discussing barriers it lists as key issues:

- Environmental barriers should make full use of current techniques and materials and be designed to solve site-specific problems;
- Widening can be an opportunity to install better environmental barriers than those presently in place and to improve the quality of life for people living close to the motorway, even though traffic may be moved nearer to them.[13]

When discussing heritage it lists as a design objective:

- To minimise noise and visual intrusion into the setting of individual buildings or sites.

While under mitigation principles it lists:

- Noise screening with appropriate environmental barriers to limit noise intrusion at source and minimise the risk of inappropriate noise insulation of historic buildings.[14]

A possible reason for at least some of the reluctance to follow the *DMRB* guidance is that all calculations of traffic noise, for environmental assessments and noise insulation purposes, must be carried out using the *Calculation of Road Traffic Noise (CRTN)* procedure.[15] This method deals with the performance of conventional reflective barriers, but has not been extended to include procedures for assessing absorptive barriers and the range of novel edge-details which is now available. The need to update *CRTN* in order to give due credit for the additional expense of including these features in a road scheme was recognised by the Highways Agency in 1996 when it commissioned a research project; however, neither a revision of *CRTN* nor the research report have been published.

A change of government in 1997 heralded a change in transport policy with the publication of *A New Deal for Roads*,[16] Notable policies included were (a) a commitment to specify quieter road surfaces for all new roads and road improvement schemes and (b) to establish a budget for noise mitigation on existing noisy roads. A relatively small budget of £5 m per year was allocated for quietening existing roads and was committed until 2011. Qualifying sites were selected by calculating the noisiest sections of road where significant numbers of residents were affected by high noise levels. While this change in policy is welcome, and brought the UK into line with many other European countries, the lack of any explicit noise reduction target in the policy has resulted in proliferation of modest structures which provide little real benefit to residents. Arguably, the budget may have been better spent by providing fewer, more ambitious and substantial barriers, which were designed to provide larger reductions in noise.

## Local government policy

Roads other than motorways and trunk roads are the responsibility of the County Councils and Unitary Authorities and, although these generally follow Government policy, they are free to adopt different policies. The use of traffic noise barriers provides a good example of the difference in approach used by the Highways Agency and local highways authorities. The practice of the Highways Agency is to consider their use where they are considered to be cost-effective, which usually means that they are used to reduce noise levels to below 68 dB $L_{A10,18hr}$, thus avoiding the cost of noise insulation. Increasingly, the cost of compensation claims under Part 1 of the Land Compensation Act has also been included in this analysis. Many local authorities go further than this and consider their use for properties exposed to lower levels of traffic noise. For example, some councils consider using barriers where properties would be exposed to more than 60 dB $L_{A10,18hr}$, while others consider their use at a level of 65 dB $L_{A10,18hr}$.

The use of the 'cost-effective' test by national Government leads to barriers being used predominantly to protect only relatively large residential areas. In contrast, many local authorities will consider their use for isolated properties and would also screen public open space and recreation areas. Schools and hospitals would be screened by a small number of local authorities, but these buildings continue to be excluded from noise insulation or financial compensation programmes under national legislation.

### Alternative European approaches

When travelling throughout Europe, it is noticeable that different countries have developed their own distinct approaches to the use of noise barriers and quiet road surfaces in order to control traffic noise. In northern European countries in particular, the use of noise barriers is much more extensive than in the UK and the scale of the barriers is generally much larger. Also of note is the extensive use of quiet road surfaces, particularly in the Netherlands.

The different approaches are a clear result of the different policies adopted by each country. The most significant of these is that noise limits, or at least objectives, are set out in legislation. Often these limits are set at low levels and limits of 55 dB $L_{Aeq}$ are frequently imposed. In the UK there is no legally defined limit, although the 68 dB $L_{AEQ,18HR}$ noise insulation threshold has become a *de facto* limit. A review of European countries' standards revealed that the objectives which these countries set themselves lie in the range 55–65 dB $L_{AEQ}$ for daytime noise exposure.[17] It should be noted that all European countries with the exception of the UK use the equivalent continuous noise level index to define their noise. That survey also revealed that many countries set night-time noise limits, whereas others simply rely on the use of a daytime limit. To a large extent the daytime limit will effectively limit night-time noise levels, since the night-time traffic volumes on most roads tend to be about 10 per cent of the total daytime flow. This assumption is not valid for those roads, which have an unusually high flow of traffic at night, for example roads serving ports or food distribution depots, and increasingly the need for night-time noise limits is being addressed.

The enforcement of standards is vital if a better environment is to be realised. This is taken particularly seriously in the Netherlands and was demonstrated during the construction of the A28 motorway at Zeist. The motorway was built adjacent to a large housing development but its opening was delayed for three years until a suitable barrier was constructed. This delay was due to the failure to comply with noise regulations and mounting public concern about this omission. This barrier eventually took the form of an extremely large, visually successful, cantilevered concrete construction which extends over the hard shoulder of the eastbound carriageway (Figures 2.1 and 2.2).

Another key difference is in the attitude towards noise from existing roads. Many countries have programmes for identifying the noisiest of their roads and implementing noise control measures. By dealing with the problem of

2.1 Substantial concrete cantilevered barrier at Zeist, the Netherlands

2.2 Massive support structure on the public side of the barrier

existing roads, rather than treating just new and altered roads, far greater numbers of people are benefiting from reduced noise levels. A striking example of this is found on the A16 at Dordrecht in the Netherlands. Here, traffic noise had increased over time and the local residents' association persuaded a member of parliament and an environment minister to sleep the night in one of the apartments to assess the noise at first hand. The officials then understood how bad the situation was and subsequently a 10 metre high, cantilevered barrier of mixed materials was constructed (Figure 2.3).

Setting low noise levels as the design objective necessarily leads to larger and more extensive barriers being required. However, in those countries which set low noise limits, the use of quiet road surfaces does much to reduce the scale of barriers used to meet the objectives and therefore produces more visually acceptable solutions.

**The Noise Insulation (Railways and Other Guided Transport Systems) Regulations, 1996**

The revival of the building of railways and tramways in the 1990s highlighted the anomaly which existed, whereby it was possible to provide noise

2.3 Large cantilevered barrier protecting a housing area at Dordrecht, the Netherlands

insulation for those dwellings badly affected by road traffic noise but there was no means of similarly treating properties affected by noise from these new rail systems. The Noise Insulation (Railways and Other Guided Transport Systems) Regulations 1996[18] were introduced to meet this need, and were specifically designed to provide the same degree of protection as the existing regulations for road traffic noise. Again there is a duty to provide noise insulation for dwellings badly affected by noise from a new or additional railway line, and also a power to carry out similar works for properties affected by noise from altered existing rail systems. There are, however, some notable differences between the two sets of regulations.

Firstly, train noise is measured using the equivalent continuous noise level index (dB $L_{Aeq}$), which acknowledges the intermittent nature of railway noise. Secondly, and more significantly, there are limits set for both daytime and night-time. There is a duty to carry out noise insulation works, or make a grant for such works, if all of the following conditions are met:

- the noise from a new or additional railway system exceeds a daytime noise level of 68 dB $L_{Aeq,18hr}$ or the night-time noise exceeds a level of 63 dB $L_{Aeq,6hr}$;
- the relevant noise level is at least a 1 dB(A) greater than the prevailing noise level;
- the noise from the railway makes an effective contribution to the relevant noise level of at least 1 dB(A).

The same minimum qualification criteria apply to the discretionary power to offer noise insulation for properties affected by noise from altered railways. The assessments are made following the calculation procedures set out in the *Calculation of Railway Noise*[19] and, in keeping with the procedures for road traffic noise, the relevant noise level is taken as the highest noise level expected to be created within 15 years of opening of the railway.

## European Union directives

Throughout its life the European Union has sought to influence only those aspects of environmental noise which may be seen as possible obstructions on the path towards a single market. Thus, its directives have been restricted to specifying maximum noise level limits for certain types of machinery, aeroplanes and road vehicles. However, in 1996 there was a notable departure from this policy with the publication of a Green Paper which addressed the need for a European noise abatement policy.[20] Significantly the Green Paper noted that the previous policy of controlling and reducing road vehicle noise emission levels had not been successful in achieving a worthwhile reduction in environmental noise exposure.

The Green Paper reviewed the noise impacts of eight major sources of environmental noise and identifies road traffic as the most important source, accounting for 90 per cent of the European Union exposure to daytime noise levels of more than 65 dB$L_{Aeq}$. Thus, some 80 million people in the European

Union lived in 'black areas' which were exposed to traffic noise at a level identified by the OECD as having a significant adverse effect on human health.[21] The OECD also defined 'grey areas' as being areas with daytime noise levels of between 55–65 dB $L_{Aeq}$ and the Green Paper noted that, although the number of people in black areas had been reduced, the number of people in grey areas had continued to rise. It was suggested that the EU adopted the following targets for the reduction of environmental noise exposure:

- phasing out of exposure above 65 dB $L_{Aeq}$ (black areas);
- reducing the proportion of the population exposed to between 55–65 dB $L_{Aeq}$ (grey areas);
- noise levels in existing quiet areas should not rise above 55 dB $L_{Aeq}$;
- exposure to more than 85 dB $L_{Aeq}$ should never be allowed.

In order to achieve these objectives, it was agreed that there should be harmonisation of measurement and prediction methods and indices, improved information exchange and publication, a common environmental assessment framework and an obligation on member states to take the necessary action to meet agreed minimum noise quality targets.

As a starting point the European Commission decided that noise exposure mapping should be carried out, either by survey or prediction, to identify both areas and populations exposed to excessive noise and the quiet areas to be preserved.[22] A study of noise indices resulted in the $L_{den}$ being adopted as the agreed harmonised index for the assessment of noise. This index is the equivalent continuous noise level measured over a 24-hour period, with the evening and night-time hours weighted by 5 and 10 dB respectively. Noise maps are now starting to be published by member states showing areas affected by transportation and industrial noise in major areas of population. These maps will allow member states to establish the number of people exposed to noise levels above 55 dB(A) $L_{den}$ and 50 dB(A) $L_{night}$ from major roads, major railways, major airports and in large urban areas.

**Risk to health**

It is beyond doubt that continued exposure to high levels of traffic noise has adverse effects on health. In 1980 the World Health Organisation (WHO) issued recommendations for the limitation of environmental noise[23] and revised guidance was published as *Guidelines for Community Noise* in 1999.[24] In this document, it is noted that 'If negative effects on sleep are to be avoided the equivalent sound pressure level should not exceed 30 dB(A) indoors for continuous noise. If the noise is not continuous, sleep disturbance correlates best with $L_{Amax}$ and effects have been observed at 45 dB(A) or less. Noise events exceeding 45 dB(A) should therefore be limited if possible'. This may be equated to external levels of 45 dB $L_{Aeq}$ and 60 dB $L_{Amax}$ based on the attenuation through a window, opened normally for ventilation, of 15 dB. These levels are well below the thresholds at which noise mitigation needs to be considered in the UK. It must be emphasised that, as its name implies, the World Health Organisation's work is solely related to health issues and, therefore, it follows that exposure to environmental noise at levels above its recommended limits must necessarily present some risk to health.

Performing tasks requiring concentration while exposed to high levels of traffic noise results in fatigue, annoyance and mistake-making. Such effects are well understood by anyone who has worked in noisy conditions; however, the long-term effects on health are less well known. Several researchers have linked long-term exposure to traffic noise with increased risk of heart disease. Ising and Michalak[25] reported that increased stress caused by noise induced communication disturbance results in changes in blood pressure, which in turn can lead to gastro-intestinal disease, hypertension and other heart and circulatory diseases. Babisch *et al.*[26–28] studied the effects of exposure to traffic noise on large samples of men in various cities and reported an increased incidence of a range of heart diseases.

## Design process

The realisation of any road or rail infrastructure project is a complex and iterative process. It will involve a wide range of specialists, and for a project to be successfully completed all of the disciplines should act as an integrated team from the project's inception. It is always a false economy to design a project with a core team of mainstream engineers and then call in various environmental specialists to improve their chosen scheme. The results of that approach are invariably inadequate, with such mitigation that can be provided often looking like the afterthought that it clearly was.

Today, any major infrastructure project will require an environmental assessment to be made and an environmental statement to be published as part of the design process. Environmental assessment should also form part of the design process on smaller schemes for which the environmental assessment process is not mandatory. The Highways Agency has set out procedures for the environmental assessment of all of its projects in the *DMRB* Volume 11.[29] This guidance requires that all aspects of the design of a road scheme should be considered throughout its planning stages, with increasing degrees of detail being required as the design develops. The process is designed to identify the effects of a scheme and to comment on their significance. Although considerable detail is given on how the effects should be quantified, no advice is given on the assessment of significance, and it has been left to individual scheme design teams to address the issue of the significance of individual effects and their cumulative effects. Official guidance on environmental assessment does not exist for non-Highways Agency road schemes nor for railways, and therefore the *DMRB* procedures are often adopted for these projects.

The nature of noise assessment is inherently different from that of the landscape assessment. The effects of noise can be quantified, whereas the landscape assessment has to be more subjective in its approach. Despite this difference, it soon becomes clear that the noise and landscape issues are, for many schemes, the most closely interrelated of all of the environmental design issues, since both deal exclusively with the effects of the scheme on people.

A difference between the two disciplines can be seen in their vocabularies. On many major schemes the landscape architect has been asked to assess the landscape mitigation as being 'essential' or 'desirable'. Such terms, however, have been used rarely by acousticians, which reveals the fact that their role has often been seen as simply the calculator of noise effects and benefits of mitigation measures, but not as a participant in the decision-making process. However, there is an increasing trend to openly define the environmental design objectives of a scheme and, therefore, all of the mitigation measures necessary to meet these objectives can rightly be described as essential.

## Acoustic assessment

At the first stage of an environmental assessment the merits of several route options may need to be evaluated. For the initial assessment the *DMRB* advice is that numbers of noise-sensitive properties should be counted in a range of distance bands alongside each route. From this it may be concluded that it is desirable to site new routes away from existing development; however, this generally means introducing noise into a wide area which may otherwise be tranquil. It may often be better to select a route within an existing noisy corridor, such that the increases in noise would be small and controllable.

During the later stages of the assessment, the precise changes in noise level at all affected noise-sensitive properties must be evaluated. A calculation of the change in the percentage of the population that is likely to be bothered by noise must also be made. There is, however, no guidance given to the designer on what emphasis should be placed on the importance of these changes. Thus, to enable sensible development of noise mitigation proposals to be made, it is essential that clear objectives be set for the project. If these are not defined by the promoter of the scheme, it is the duty of the design team to propose and agree suitable targets. If this is not done, any noise mitigation which may be provided will be included on an *ad hoc* basis.

Different design objectives will be appropriate to different types of scheme; for example, a new road in a quiet greenfield site cannot reasonably be built without some increase in noise, but when widening an already noisy motorway it is quite feasible to design the scheme to reduce noise levels for local residents. It is not possible, therefore, for a single design objective to be set for all schemes, but it is desirable that similar objectives are set for schemes of a similar nature.

Objectives can be set in terms of absolute levels (levels not to exceed 68 dB $L_{A10,18h}$), relative levels (allow no more than 5 dB increase in noise) or in terms of the significance criteria set for the assessment of the scheme. The significance criteria are used as part of the final assessment of the scheme, but must be known during the design stage in order that the relative importance of the various environmental effects of the scheme can be properly evaluated. Significance criteria are defined using descriptors such as minor, moderate and major to describe both the positive and negative effects of the scheme and may take account of the nature of the area affected by these changes. Thus, an increase in noise of say 10 dB may be described as moderately adverse if only a few isolated properties are affected, but would be seen as a major adverse effect if a large community was affected. If using significance criteria to set design objectives, it may be reasonable to avoid creating any effects worse than moderately adverse for a new road, whereas when dealing with alterations to existing roads the objective may be to provide at least minor beneficial effects.

Where projected noise levels exceed the design objectives, it will be necessary to evaluate all the noise control options available. These include the use of low-noise road surfaces, lowering the road to utilise the screening of natural topography, the use of barriers and in some rare cases the use of speed restrictions. Often a combination of measures will be appropriate in order to meet both the acoustic requirements and the needs of other members of the design team. Arguably any solution required to meet the design objectives can be described as essential, but these must be balanced against their impact on other issues, such as their visual impact, and due regard must be paid to the cost of the mitigation.

This latter point is seldom addressed properly. All too often a cost–benefit analysis is carried out which simply weighs the cost of the mitigation against the alternative cost of providing noise insulation and, increasingly, the cost of compensation for loss of property value. Such evaluations may be well-intentioned but they are essentially flawed, since they place no value on the often considerable social and public health costs that increased noise exposure incur. This is particularly surprising since the hypothetical costs of delays caused to the travelling public are included in the cost–benefit analysis of a scheme, both when addressing congestion on existing roads and the likely delays which would be caused during the construction process.

In order that the noise effects of a scheme, with its proposed mitigation, can be balanced against the other issues, it is necessary to define the significance of any changes. These significance criteria must be developed as part of an overall assessment framework for the scheme and the descriptors must be consistent across all disciplines involved in the assessment. In general, the significance criteria should be similar for all projects, but they must also reflect local needs.

## Landscape decision-making process

### Landscape assessment

In the UK, the Department for Communities and Local Government's *Planning Policy Guidance 24: Planning and Noise (PPG 24)* states that 'noise can have a significant effect on the environment and on the quality of life enjoyed by individuals and communities'.[30] This demonstrates the UK government's commitment to protect and maintain the character and quality of designated areas. However, appropriate action is also required in areas where the current and future landscape character, quality and values are and would be diminished and where important aspects of the landscape, such as 'genius loci' (sense of place), uniqueness and tranquillity are being eroded.

It is the landscape architect's role to try to minimise the effects of any scheme on the landscape. Where adverse effects occur, mitigation must be designed. Landscape objectives must be formulated to satisfy this national requirement. Local needs and issues also need to be considered.

The main landscape objective of a road or railway scheme should be to avoid any detrimental change in landscape character as this defines the nature of any particular location. Secondly, deterioration in landscape quality or visual intrusion must be avoided.[31] Where a scheme is proposed in an already downgraded location, opportunities should be taken to improve the environment. The environmental assessment process requires that the significance of the effects of the scheme be identified. To do this, significance criteria are developed to reflect the local landscape conditions and take into account the scale of effects. Throughout the design process these criteria are used to evaluate and inform the design. The factor of time is an important difference between landscape and most other environmental interventions: most large-scale mitigation will use immature plants, which will only meet the design objectives once they have reached a reasonable state of maturity.

When dealing with landscape issues, the landscape architect has to use judgement to define and assess certain aesthetic qualities. This leads to comparatively subjective assessments and thus 'subjective' design and mitigation proposals, notwithstanding the fact that these may be based on the well-thought-out and rational principles of an experienced professional.

**Assessment of visual intrusion**

The assessment of visual intrusion must relate to people, as it is human beings who are affected by unsightly views. People who are exposed to views are described as receptors. Visual intrusion relates directly to the distance of the receptor from the source of the visual disturbance. It must also relate to the magnitude of the visual disturbance and how this affects the quality of the view. Account must be taken of the number of people affected and their attitude towards the intrusion, thereby informing the degree of significance of the change in view.

As part of the assessment and design process, visual intrusion and Zone of Visual Influence plans (ZVIs) are drawn up. These are used with the landscape character plans to help determine the type of barrier that would be appropriate. An example of where a barrier design meets these visual requirements is found in Bern, Switzerland. Here, a viaduct crosses high above the River Aare. Most views to the viaduct are from the surrounding hillsides and the valley below. The strategic nature of the views has been acknowledged by varying the use of barrier materials relative to the visibility of the traffic on the viaduct. Where views of the traffic are less intrusive, a barrier of transparent acrylic sheeting is employed, thereby maintaining the integrity of the form of the viaduct. On the other hand, where views of the traffic are deemed to be undesirable, the barrier is made from opaque perforated aluminium. An additional advantage of this solution is that the transparent sections also allow drivers views out from the road corridor (Figures 2.4–2.6). Visual intrusion will vary with time and plans must reflect

**2.4** Viaduct with transparent and opaque noise barriers

**2.5** Transparent screen on a viaduct in a zone of low visual intrusion

**2.6** Opaque screen on a viaduct in a zone of high visual intrusion

this. A typical plan may show the changes in the magnitude of visual intrusion at three stages; existing situation, the situation at the year of opening and that 15 years on, when the vegetation will have become established (Figure 2.7).

Another useful tool for determining the use of the correct forms, colours and textures is to produce colour, form and texture montages of the landscape within which the barrier is to be located. This allows the designer to decide whether to fit the barrier into its visual context or to try to make a more pronounced statement by contrasting forms, colours and textures.

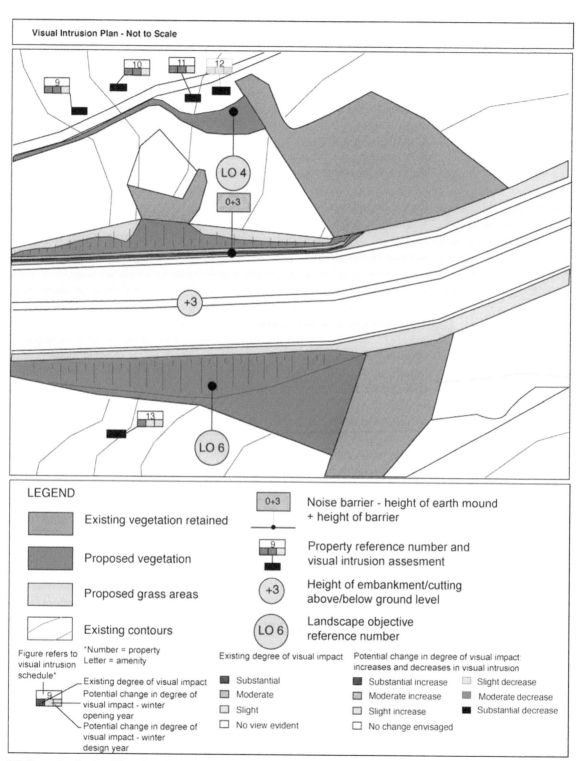

**Visual Intrusion Plan - Not to Scale**

LO 4

0+3

+3

LO 6

LEGEND

Existing vegetation retained

Proposed vegetation

Proposed grass areas

Existing contours

Figure refs to visual intrusion schedule*

Existing degree of visual impact

Potential change in degree of visual impact - winter opening year

Potential change in degree of visual impact - winter design year

0+3 — Noise barrier - height of earth mound + height of barrier

9 — Property reference number and visual intrusion assesment

+3 — Height of embankment/cutting above/below ground level

LO 6 — Landscape objective reference number

*Number = property Letter = amenity

Existing degree of visual impact

- ■ Substantial
- ▨ Moderate
- ▢ Slight
- □ No view evident

Potential change in degree of visual impact: increases and decreases in visual intrusion

- ■ Substantial increase
- ▨ Moderate increase
- ▢ Slight increase
- □ No change envisaged
- ▢ Slight decrease
- ▨ Moderate decrease
- ■ Substantial decrease

**2.7** Example of a visual intrusion plan

## *DMRB* assessment, consultation and design process

The Highways Agency sets out its design process in the *DMRB*, 'Stages in the Development of the Preferred Solution', under the heading the 'Environmental Barrier Design Process'.[32] The steps described to reach the preferred solution are laid out in a simplified form and in reality the process is more complex and iterative. The description of the procedure also does not mention the involvement of local government and other interest groups who are usually consulted at an early stage. Furthermore, the public who may be affected by the proposals are not mentioned since they are consulted only after the preferred solution is announced. This shows the difference between planning strategies used in the UK and those overseas.

The *DMRB* barrier design process is outlined below in stages A to K. Comments in braces amplify the process which takes place, indicate potential improvements which should be made and notes where public involvement has been found to be beneficial in other countries. Figure 2.8 illustrates this procedure.

### A. Consider initial alignment options

Investigate potential routes in order to minimise adverse environmental impact of the new road. {This includes initial acoustic, landscape and environmental surveys and assessments.}

### B. Identify affected communities and areas

Highlight communities, facilities, recreation areas and designated areas alongside the route potentially affected by noise and visual intrusion. {This includes potential effects on landscape character and landscape quality.}

### C. Review alignment options

Investigate modifications to vertical and horizontal alignments, in order to reduce the impact of the road in terms of noise and visual intrusion. {Stages A, B and C are an integrated process and may be repeated several times to accommodate the priorities of different disciplines and also incorporate the results of surveys as they become available. Local government would also be consulted.}

### D. Identify noise reduction and visual screening objectives for each location

Determine location(s) and height(s) of barriers required to achieve the target reductions and establish the most effective profile providing an acceptable level of protection.

Confirm the need for a barrier before proceeding further. {Objectives should be set at the start of the process. If barriers are not required proceed no further. Although the locations and heights of barriers can now be defined for acoustic reasons, a knowledge of Stage E is required

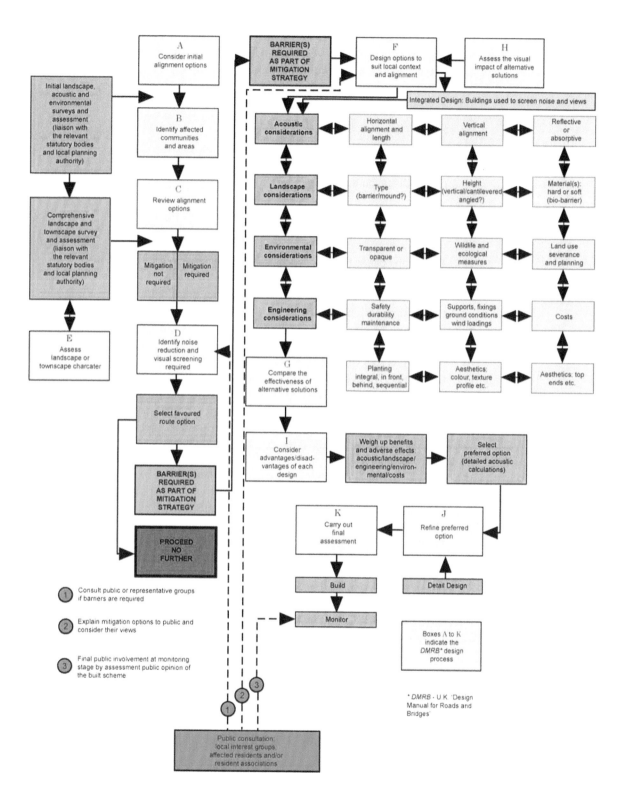

**2.8** Mitigation and barrier selection process including stages of public involvement

for the determination of landscape barriers. Consider other solutions including noise absorptive pavements and insulation. If barriers are required, local interest groups and residents should be consulted.}

E.  Assess landscape or townscape character

Identify the main features of the locality which could influence the range of barrier solutions considered, drawing on the landscape assessment for the route.

{Select the favoured route option}

{Stages A to E identify the possible routes and their mitigation. The least damaging route option is now chosen paying due regard to cost.}

F.  Design options to suit local context and alignment

Decide on the form of the barrier (earth mounding, fence, wall, structure or proprietary system etc.) which would be most compatible with the neighbourhood. Select the most appropriate materials for the protected side compatible with the landscape or townscape character of the neighbourhood. {A barrier has two sides. Therefore, also select the material for the traffic side of the barrier. An overall strategic design concept is also required which would provide an overall identity for the route. However, there may be conflicting objectives to be resolved between disciplines. Compromises may be required. The public should be informed about the possible mitigation options and their views should be considered.}

G.  Compare the effectiveness of alternative solutions

Consider whether there is a case for using noise-absorbing or -dispersing surfaces to reduce noise reflected from the barrier. Confirm whether the target reductions in noise would be acceptable. {Consideration of noise absorptive and other mitigation types should have been considered at Stage D.}

H.  Assess the visual impact of alternative solutions

Clarify the visual impact of alternative designs on affected residential or other sensitive areas using two- or three-dimensional sketches. Consider the use of planting to reduce the visual intrusion of the barrier itself. Consider the use of transparent materials to reduce adverse impacts such as loss of views or light. Confirm whether the target reductions in visual intrusion would be achieved. Should the barrier have the same appearance on the road user side as that selected for the protected side. {All of the above should have been done at Stage F.}

I.  Consider advantages/disadvantages for each design

Compare the characteristics of options, including implementation and maintenance costs, to inform choice of preferred option.

{Select the preferred option}

{The most appropriate solution for each barrier is now selected.}

J.  Refine preferred option

At the detailed design stage refine preferred solution to optimise visual and noise benefits. Consider visual impact on road user, including: monotony – the need to provide drivers with visual relief – harmonisation of lighting, street furniture, signs, etc. {These latter issues should have been done at Stage F as they inform the choice of barrier. Barrier specifications should be written to ensure that the design objectives are fully realised.}

K.  Carry out final assessment

Ensure all relevant criteria have been met.

{Build}

{Supervision should ensure that the design objectives are realised.}

{Monitor}

{The appropriate maintenance and monitoring should be carried out at appropriate intervals after construction to ensure that the barrier and other mitigation measures are performing and that landscape measures develop as intended. The public should also be surveyed to assess their opinion of the scheme to inform future designs.}

## Summary of mitigation design strategies

In order to comply with the government's guidance on the design of new transport infrastructure projects and improvements to existing infrastructure, a scheme must include the appropriate mitigation. This must be designed to minimise the impact of the scheme and, where this includes barriers, their impact must be minimised. Mitigation strategies will depend on the landscape character of the area and will vary for rural and urban locations. A summary of the ways to achieve this is given below, starting with the most effective (Figure 2.9).

### Rural and semirural locations

1.  Consider locating the route away from sensitive areas to avoid the need for barriers.
2.  Where possible, place the corridor in tunnel where its effects are contained. (This may be deemed to be too expensive in many rural locations.)
3.  Contain the road within a cutting to provide acoustic and visual screening.

1. Distance

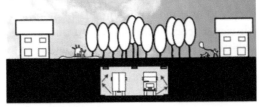

2. Tunnel / cut and cover

3. Cutting

4. False cuttings / earth mounds

5. Barriers: bio-barriers/vertical/cantilevered/angled

6. Combined solution

7. Quiet surfacing: porous asphalt / whisper concrete

8. Insulation

**2.9** Noise mitigation options

4.  Should a barrier be required, select natural structures, such as earth mounds with appropriate planting.
5.  Where space is limited, consider 'bio-barriers' to achieve a natural effect with planting on adjacent land if appropriate.
6.  Vertical or cantilevered structures with planting on either side.
7.  Where planting cannot be achieved, the materials and detailing should be of a very high standard. If views need to be maintained then transparent barriers need to be considered. In most cases transparent barriers should be used on bridges and viaducts.

Urban and developed locations

1.  The corridor should be placed in a tunnel where its effects are contained. This can be cost-effective in densely populated areas as the roof of the tunnel may be used for development or recreational purposes.
2.  Contain the road with non-noise-sensitive buildings which screen the noise.
3.  Contain the road within a retained cutting to minimise noise.
4.  Apply noise absorptive cladding to retaining walls.
5.  Where barriers are required, close inspection should be made of the urban fabric, materials, historical and cultural context, colour and textural patterns to allow the barrier design to tie into the local context. The location should be considered and a decision made as to whether to make the barrier an architectural feature or to blend it into the surroundings. In both cases planting can be used to advantage. Where appropriate, bio-barriers should be considered – they can be used as urban hedges. In busy urban locations the materials and detailing should be of a very high standard. If views need to be maintained, transparent barriers should be considered.

In both rural and urban locations, low-noise pavements should be considered at each stage as part of the solution. Noise insulation can be used in conjunction with a barrier to minimise its visual impact.

**Taking design strategies further**

Creating holistic and well-formed design strategies for new and upgraded roads is a forte of the Netherlands. One of the main objectives of the 'A12 Rainbow Route', which runs from Den Haag in the west, past Utrecht and then Arnheim in the east, is to 'develop the A12 into a motorway that complements the characteristics of the areas through which it passes'.[33] A comprehensive strategy develops objectives to create a more cohesive appearance in areas with particular identities, which should remove visual clutter and relentless uniformity. Environmental noise barriers, structures and viaducts will thus be designed to create visual continuity, which creates an improved visual and safer environment for motorists, as well as reducing design, construction and maintenance costs. The areas through which the route passes are divided into landscape and townscape character zones, which include meadow, urban, woodland, business and mixed residential areas. The idea is that this characterisation will help to define distinctive design parameters and thus lessen 'the amount of dull-looking business parks' and other nondescript development and 'proliferate impressive views from the motorway itself'.[34]

This strategy, which could provide a model for future strategies around the world, emphasises the integrity of context and local conditions. All elements of the motorway, which include junctions, lighting, bridges and viaducts, and environmental noise barriers, will be designed to complement the areas they pass through. Although the environmental noise barriers used in different locations stem from the same model to reduce construction and maintenance costs, modifications have been made to ensure that they integrate better into the different kinds of landscape in which they are located.

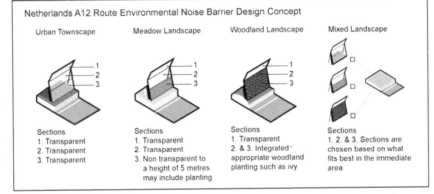

**2.10** Design of environmental noise barriers corresponding to dominant character areas. Information courtesy of Steunpunt Routeontwerp, Den Haag

**2.11** Road side façade of a barrier with seven modular panels along the A16, Rotterdam, the Netherlands

**2.12** Community side façade of the barrier showing support columns and modular panels along the A16, Rotterdam, the Netherlands

**2.13** Modular barrier used for air pollution testing at Strandnulde in the Netherlands

These include the woodland, meadow, urban or mixed landscape areas noted above (Figure 2.10).

On the A12, the environmental noise barriers all share a basic design form, based on the modular noise barrier developed in the Netherlands (discussed in Chapter 4) (Figures 2.11–2.13).

The environmental noise barriers follow the following guidelines:

- The barrier panels are the same construction and size (1 metre tall and 6 metres in length);
- The barriers are angled 10 degrees away from the motorway to a height of 5 metres. Above 5 metres the barrier is reverse-angled 10 degrees towards the motorway.

The design of the barriers varies according to the type of landscape it passes through:

- In urban areas, barriers are transparent;
- In meadow areas barriers are opaque to a height of 5 metres with additional suitable site-sensitive planting and then transparent above;
- In woodland areas, the lower half of barriers are opaque with additional suitable site-sensitive planting and then transparent above;
- In mixed areas, barriers are chosen from the three types according to the best fit in the immediate area (Figure 2.10).

## References

1. *The Land Compensation Act, 1973*, HMSO, London.
2. *Building and Buildings. The Noise Insulation Regulations*, HMSO, London.
3. Department of the Environment (1988) *Statutory Instruments, 1988 No. 2000, Building and Buildings, The Noise Insulation (Amendment) Regulations*, HMSO, London.
4. The Scottish Office (1975) *Statutory Instruments, 1975 No 460 (S.60), Building and Buildings. The Noise Insulation (Scotland) Regulations*, HMSO, London.
5. Department of the Environment for Northern Ireland (1995) *Statutory Rules of Northern Ireland, 1995 No 409, Land, The Noise Insulation Regulations (Northern Ireland)*, HMSO, London.
6. Secretary of State for Environment *et al.* (1990) *This Common Inheritance*, HMSO, London, p. 212.
7. Department of Transport (1980) *The Highways Act, 1980*, HMSO, London.
8. Secretary of State for Environment *et al.* (1990) *This Common Inheritance*, HMSO, London, pp. 208–14.
9. Press Notice No. 204, 28 July 1992, The Department of Transport, London.
10. Press Notice No. HA 168/96, 22 May 1996, Highways Agency, London.
11. The Highways Agency (1992) *Design Manual for Roads and Bridges*, Volume 10, *Environmental Design* Section 5, HMSO, London.
12. The Highways Agency (2001) *Design Manual for Roads and Bridges*, Volume 10, *Environmental Design* Section 2 Part 1, HMSO, London.
13. The Highways Agency (2001) *Design Manual for Roads and Bridges*, Volume 10, *Environmental Design* Section 2 Part 1, Annex 4, HMSO, London.
14. The Highways Agency (1992) *Design Manual for Roads and Bridges*, Volume 10 *Environmental Design* Section 1 Part 6, HMSO, London.
15. Department of Transport, Welsh Office (1988) *Calculation of Road Traffic Noise*, HMSO, London.
16. Department for the Environment, Transport and the Regions (1998) *A New Deal for Roads*, HMSO, London.

17. English, C. (1993) *Strategies for Controlling Road Traffic Noise*, Proc. Transport Noise in the 90s, London.

18. Department of Transport (1995) *Statutory Instruments, No.1996/428, Building and Buildings. Transport. The Noise Insulation (Railways and Other Guided Transport Systems) Regulations*, HMSO, London.

19. Department of Transport, Welsh Office (1995) *Calculation of Railway Noise*, HMSO, London.

20. European Commission (1990) *Green Paper on Future Noise Policy*, COM (96) 540 final, European Commission.

21. Organisation for Economic Co-operation and Development (1991) *Fighting Noise in the 1990s*, OECD, Paris.

22. The European Parliament and The Council of The European Union (2002), Directive 2002/49/EC of the European Parliament and of the Council of 25 June 2002 relating to the assessment and management of environmental noise, *Official Journal of The European Communities*.

23. World Health Organisation (1980) *Environmental Health Criteria, 12, Noise*, WHO, Geneva.

24. World Health Organisation (1999) *Guidelines for Community Noise*, WHO, Geneva.

25. Ising, H. & Michalak, R. (1998) *Stress reactions due to noise-induced communication disturbance compared with direct vegetative noise effects*, private communication.

26. Babisch, W., Ising, H., Gallacher, J.E.J. and Elwood, P.C. (1988) 'Traffic noise and cardiovascular risk. The Caerphilly study, first phase. Outdoor noise levels and risk factors', *Archives of Environmental Health*, 43(6) 407–414.

27. Babisch, W., Ising, H., Kruppa, B. and Wiens, D. (1994) 'The incidence of myocardial infarction and its relation to road traffic noise – the Berlin case-control studies', *Environmental Health*, 20(4) 469–474.

28. Babisch, W., Ising, H., Elwood, P.C., Sharp, D.S. and Bainton, D. (1994) *Archives of Environmental Health*, 48(6) 406–413.

29. The Highways Agency (1993) *Design Manual for Roads and Bridges, Volume 11 Environmental Assessment*, HMSO, London.

30. Department of Communities and Local Government (1994) *PPG 24: Planning and Noise*, HMSO, London.

31. Landscape Quality refers to an evaluation or scale applied to different character areas with regard to their character, condition and aesthetic appeal and attaching a value or grade to these areas. Landscape Value refers to the value or importance of the landscape internationally, nationally, regionally or to the local community. A hierarchy of Landscape Quality/Value may include: High Quality/Value, Moderate Quality/Value, Poor Quality/Value.

32. The Highways Agency (1994) *The Design Manual for Roads and Bridges, Volume 10, Environmental Design Section 5, Environmental Barriers, Part 1, Design Guide for Environmental Barriers*, July 1994, HMSO, London.

33. 'A12 Regenboogroute Route', Projectbureau Regenboogroute A12, Delft. (www.Regenboogroute.nl).

34. Ibid.

# 3

# Acoustic performance of barriers

## Introduction

All too often noise barriers are built which provide little or no protection to the communities that they are intended to serve. There are others which, with a little more care in the design, could have provided significantly better screening than they achieve. To avoid these costly mistakes and to ensure that the greatest possible benefits are realised from every noise barrier, it is essential for designers to understand the basic principles of acoustic barrier theory.

Sound propagates from a source as a series of rapidly fluctuating pressure waves which expand spherically as they move away from the source (Figure 3.1a). These pressure waves create the sensation of noise when they reach the listener's ear. Although sound travels as waves it is often convenient to model sound propagation as straight lines or rays, which reach the listener or receiver either directly or indirectly after being reflected or diffracted by intervening surfaces (Figure 3.1b). The ray tracing method is used below to describe commonly occurring screening scenarios.

## Barrier theory

For an unscreened road, the most important sound transmission path is the ray travelling directly between the road and the receiver, known as the direct sound, $L_{p,dir}$. Another ray that will reach the receiver is the ray which strikes the ground and is reflected upwards to the receiver, $L_{p,grd}$, (Figure 3.2). There is a degree of destructive interference between these two rays which results in greater attenuation of the $L_{p,dir}$ than would be expected by geometrical spreading alone. The precise mechanism of this attenuation is not fully understood, but it is at its greatest where the propagation is over acoustically

**3.1** Sound propagation: (a) spherical spreading of sound; (b) ray model

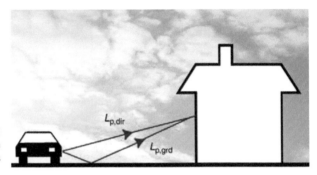

**3.2** Unobstructed sound transmission paths

'soft' ground, such as grassland, and where the $L_{p,dir}$ is particularly close to the ground. This ground attenuation is frequency-dependent and Hutchins et al.[12] showed that the destructive interference predominantly occurs in a frequency range centred on 500 Hz (Figure 3.3).

The introduction of a barrier can greatly reduce the strength of the direct ray, although for most practical barriers this will remain a potential transmission path, $L_{p,trans}$. The important ray is now that diffracted downwards from the top edge of the barrier, $L_{p,diff}$, (Figure 3.4). The presence of the barrier also eliminates the $L_{p,grd}$ as a significant sound transmission path.

Considerable work has been carried out, using ray tracing techniques, to establish the acoustic performance of a barrier, namely the difference between the $L_{p,dir}$ and the $L_{p,diff}$. Probably the most influential work is that of Maekawa[3] and this remains the basis of most of the practical methods for calculating barrier performance. Other workers developed the basic barrier theory and correlated the results with fieldwork.[4,5] The theory developed

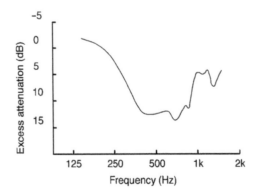

**3.3** Excess attenuation for propagation over 500 metres of soft ground

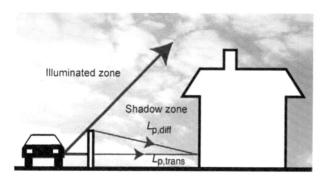

**3.4** Key sound transmission for screened noise source

calculates the acoustic performance of a vertical screen in terms of the Fresnel number $N$, which is defined as:

$$N = 2\frac{\delta}{\lambda} \qquad (1)$$

where $\delta$ is the path length difference (diffracted path length minus direct path length) and $\lambda$ is the wavelength of sound in air. In the shadow zone, the area where the barrier breaks line of sight between the source and the receiver, $\delta$ is defined as positive, and for rays propagating above the diffracting edge of the barrier into the illuminated zone, $\delta$ is negative. Kurze and Anderson[5] gave the following equation for the insertion loss $IL$ of a barrier:

$$IL = 5dB + 20 \log \frac{\sqrt{2\pi N}}{\tanh \sqrt{2\pi N}} \qquad \text{for } -0.2 < N < 12.5$$

$$= 24 \text{ dB, for } \quad N > 12.5 \qquad (2)$$

The above formula only applies to single vehicles at their closest point to the receiver and more complex expressions are available to describe the performance of a barrier for a stream of traffic. Fortunately for designers of barriers, they will seldom need to use these as the approved calculation methods for traffic noise provide the results in graphical or tabular form and computer programs are usually available which implement these given methods. An understanding of the implications of the theory is necessary, however, if the optimum benefits are to be obtained from barriers.

Firstly, from Equation (2) it can be seen that when a barrier just breaks the line of sight between the noise source and the receiver there is a 5 dB attenuation of noise, and there may be some reduction of noise for receivers in the illuminated zone. The barrier attenuation used in the UK's traffic noise calculation method[6] is shown in Figure 3.5. This gives a theoretical limit for barrier attenuation of about 20 dB(A) in the shadow zone; however, the required values of $\delta$ can seldom be realised and, in practice, a realistic limit is about 15 dB(A). The graph also shows that in the illuminated zone the barrier attenuation rapidly tends to zero at $\delta = -0.6$ and therefore there is little practical screening benefit to be gained in this region.

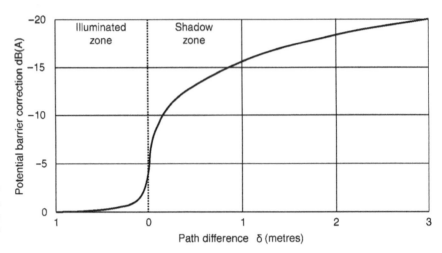

**3.5** Potential barrier correction as a function of path difference

In the shadow zone, the difference in attenuation can be some 3 dB per octave for $\delta = 0.5$ m, but to avoid cumbersome, frequency-based computation, most calculation methods have adopted a composite value of $N$. These are derived from the known frequency spectrum of traffic and the acoustic performance of a barrier, and typical composite $N$ values correspond to frequency range 300–500 Hz. Using a composite value of $N$ allows the barrier calculations to be carried out in terms of A-weighted sound levels (dB(A)).

From Equation (1), it can be seen that the acoustic performance of a barrier is frequency-dependent. It should therefore be remembered that this, together with the loss of the mid-frequency ground attenuation, will always have the effect of biasing the received sound towards the low frequency end of the spectrum when a barrier is introduced.

## Sound insulation

The attenuation achieved by a thin barrier can be compromised if it is not designed to ensure that the transmitted sound path does not significantly influence the overall noise level at the receiver. A contribution of 0.5 dB(A) to the overall level is commonly adopted as the limit for the contribution of the transmitted sound and this requires limiting the transmitted path as follows:

$$L_{p,trans} = L_{p,diff} + 10 \text{ dB} \tag{3}$$

The sound insulation provided by a barrier is dependent upon many factors such as surface mass, stiffness, loss factors and the angle of incidence of the sound. The most significant of these is the surface mass of the barrier and many calculation methods adopt this as the sole descriptor of a barrier's sound insulation. In the UK the Department of Transport[7] gives the following formula for calculating the minimum required surface mass for a barrier:

$$M = 3 + anti \log\left(\frac{A - 10}{14}\right) \text{kg/m}^2 \qquad\qquad (4)$$

where $A$ is the potential attenuation in dB(A) of the barrier $(L_{p,dir} - L_{p,diff})$.

Only the surface mass of the barrier's panels should be considered in this equation and the mass of any support posts, etc. should be disregarded. It is also essential that there are no gaps in the barrier to avoid leakage of sound. For large gaps the sound waves will pass through unattenuated; however, in the case of narrow slots the incident sound may actually be amplified.

## Barrier placement

To obtain the optimum performance from a barrier it is generally desirable to place it as close to the road as possible. This applies where the road and the receiver are level, or where the road is elevated on an embankment or a viaduct (Figure 3.6). The same performance can be obtained by placing the barrier close to the receiver, but this is seldom practical and can only be considered for isolated properties some distance from the road.

The conventional wisdom that a barrier should be placed as close as possible to the source or the receiver is not valid where the road is in a cutting, or where a raised landform separates them. Here, the barrier is best placed at the top of the cutting slope (Figure 3.7). It must be recognised that barrier placement presents a potential conflict between the acoustic and visual objectives, and a compromise may need to be reached.

**3.6** Barrier should be placed as close to the road as possible on flat or elevated ground

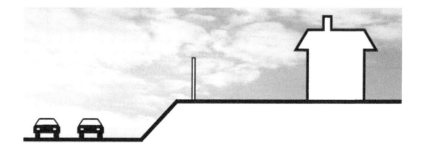

**3.7** Barrier should be placed at the top of the cutting slope

**3.8** Central reservation barrier combined with side barriers at Hardinxveld, the Netherlands

For any barrier the path difference will be progressively less for each traffic lane as their distance from the barrier increases. Thus, for an area protected by a barrier, it is usually the furthest lane of traffic which dominates the noise at the receiver. Simply increasing the height of the barrier will not alter the dominance of the noise from the furthest lane of traffic and can lead to the use of unacceptably high barriers. In these situations the use of a second barrier, located between the carriageways, can be of considerable advantage, since the two barriers are kept as close as possible to the two sources of noise (Figure 3.8). This technique allows the overall height of the barriers to be minimised and is of particular value on (a) roads where the outer edges of the carriageways are widely separated, such as dual three- or four-lane roads or dual two-lane roads with abnormally wide central reserves and (b) situations where there are receivers located above the road level.

## Barrier length

Diffraction of sound occurs not only at the top edge of a barrier but also around the ends. Thus, the overall noise reduction provided by a barrier depends not only on its height and location between the noise source and the receiver but also on its length. The sound diffracted around the ends of a barrier will tend to be less significant than the sound diffracted at the top edge, as this transmission path will still benefit from any ground absorption effects which would have attenuated the direct ray. It is generally found that a barrier covering an angle of 160° subtended from the receiver to the road will ensure that the end diffracted rays are not significant (Figure 3.9). The length of the barrier can be reduced by angling the ends away from the road if space is available (Figure 3.10).

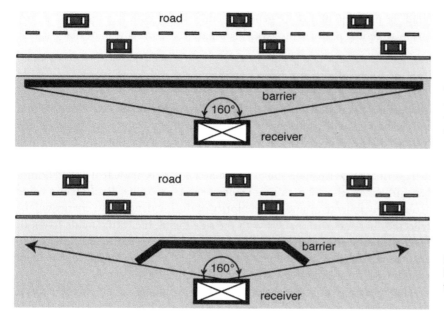

**3.9** Minimum angle of view which should be screened to avoid barrier being short-circuited by transmission around ends

**3.10** Barrier length reduced by angling ends away from the road

## Bunds versus screens

Visual considerations, space or the availability of spoil will often dictate that earth mounds or bunds are used as barriers, particularly in rural areas. Conventional wisdom has held that bunds and screens of equal height provide the same degree of attenuation, and indeed the CRTN calculation method does not distinguish between the performance of these two barriers. It is important to note that, in general, the height of a bund will need to be greater than a vertical screen, since the screening edge cannot be placed as close to the noise source. The particular geometry of each screening situation will determine the increase in height needed for a bund, but typically this may be in the order of 0.5–1 m.

Theoretical studies have indicated that vertical screens may be more efficient barriers than bunds. Using a boundary element model, Hothersall et al.[8] reported that, for traffic noise, a vertical screen gave 4 dB greater attenuation than a wedge-shaped bund of similar height. This difference was reduced to 0.5–1 dB when a flat-topped bund was studied.

In a scale model test using a vertical screen mounted on top of a bund, Hutchins et al.[1,2] showed an excess attenuation of 10 dB over a relatively narrow frequency range (350–500 Hz). This was considered to be due to destructive interference at the top of the barrier between the direct ray and the ray reflected off the sloping face of the bund. This can only occur where the direct ray is parallel to the slope of the bund, that is, when the traffic is close to the barrier and therefore it is only effective for the closest traffic lanes. The research did not report overall dB(A) improvements for multilane situations, but these may be expected to be slight.

From this it may be concluded that, where bunds are to be used, their performance may be optimised by the use of a low vertical screen on the top of the bund. In practice, bunds are generally constructed with flat tops and so any loss in screening performance when compared to a screen will be marginal.

## Dealing with reflections from barriers

The earlier discussion of barrier theory showed that the introduction of a simple barrier will significantly reduce the direct sound ray between the source and the receiver ($L_{p,dir}$). The sound energy falling onto the barrier will either be reflected or absorbed by the barrier, with only a small proportion being transmitted through the barrier. Barrier panels made from solid homogeneous materials, such as timber or concrete, may be described as being acoustically reflective, since the majority of the incident sound energy will be reflected. Barriers with panels with the side facing the traffic made from, or filled with, a porous material, such as mineral fibre or foam, are classified as absorptive barriers since these will absorb the majority of the incident sound, with only a relatively small proportion being reflected.

The reflected ray can be considered to originate at an image source, located at the same distance as the source from the barrier, but on the far side of the barrier. Although this image source is further from the receiver than the true source, it will always be subject to less screening than the primary diffracted ray, and often there will be no screening of this image source ray $L_{p,img}$ (Figure 3.11). This can seriously degrade the attenuation provided by the nearside barrier. For 2 metre high parallel reflective barriers 34 m apart Watts[9] reports a 4 dB(A) loss in performance of the single nearside barrier. This loss in performance is greater as the barrier heights increase or the separation distance is reduced, and for 4.5 m parallel reflective barriers 18 m apart, a loss in attenuation of 6 dB(A) may be expected.[10]

To overcome this problem of reflected noise, it is possible to simply increase the height of the barriers. This can, however, substantially increase the cost of the barriers and also result in visually unacceptable effects. A common solution is to tilt the barriers outwards from the road in order to reflect the noise above the receiver rather than towards it (Figure 3.12). The angle of

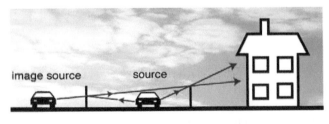

image source          source

**3.11** Image source model for reflected sound: windows are screened from direct ray but not from image source ray

**3.12** Angled barriers reflect sound upwards

tilt required depends on the barrier separation. Slutsky and Bertoni,[11] showed that for a 45 m separation, a tilt angle of only 3° was required to allow the full screening potential of the nearside barrier to be realised; however, for an 18 m separation, tilt angles of 10–15° were required.

An alternative solution to the problem of reflection is to use barriers with sound-absorbing surfaces facing the traffic. These ensure that the majority of the sound energy falling onto the barrier is absorbed and that only an insignificant reflected ray is produced (Figure 3.13).

There are two principal classes of sound-absorbing mechanisms which are utilised in absorptive noise barriers. The majority of absorptive barriers incorporate a permeable layer of material which faces the incident noise. The flow resistivity of this porous material causes the acoustic energy of the sound waves to be dissipated within the material and eventually transformed into heat energy. The second class of absorber is based on the Helmholtz resonator principle, whereby the incident sound wave enters a series of cavities in the barrier via small holes or narrow slots. These systems are only effective at discrete frequencies, related to the depth of the cavities. The range of frequencies over which absorption can be provided may be significantly widened by the introduction of a layer of porous sound-absorbing material into the cavity. This selective performance is not necessarily a disadvantage, however, since it can be used to offset the biasing of the sound spectrum towards the lower frequencies as a result of introducing a screen.

The degree of sound absorption provided by a cavity absorber is controlled by the ratio of the area of the inlet opening and the cross-sectional area of the cavity. Although it may be desirable to maximise this ratio, it is likely to be counterproductive, as it would result in a predominantly reflective outer surface.

A further option for controlling reflected sound is the use of barriers with a profile which disperses sound, i.e. reflects sound in a random manner. However, while the benefits may be significant for a single vehicle, the cumulative effects of the dispersed sound from a stream of vehicles results in very little overall improvement when compared with flat reflective screens. May and Osman[12] reported that a dispersive barrier with a vertical corrugated profile gave only 0.5–1.1 dB improvements when considering a six-lane road. The decision to use this type of barrier is likely to be made on structural and cost grounds, since they can be inherently more stable than a flat vertical or inclined barrier.

## Reflections from vehicles

Reflections of sound between the sides of vehicles, particularly high-sided vehicles, can also significantly reduce the acoustic performance of a barrier.

**3.13** Facing a barrier with a sound absorptive material greatly reduces the strength of the reflected ray

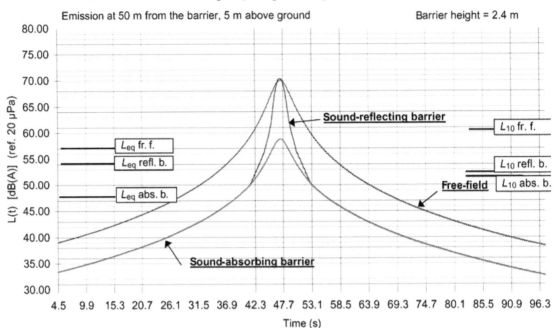

Level / Time evolution during the passage of a lorry at 100 km/hr in the first lane

Emission at 50 m from the barrier, 5 m above ground                    Barrier height = 2.4 m

**3.14** Effect of reflective and absorptive barriers on single vehicle pass-by. The graph shows the A-weighted sound pressure level (relative to the standard reference pressure) during a single vehicle pass-by. The equivalent continuous sound levels ($L_{eq}$) for the pass-by are shown on the left field (fr.f.), i.e. without a barrier, with a reflective barrier (refl.b.) and with an absorptive barrier (abs.b.). The corresponding A-weighted sound pressure levels exceeded for 10 per cent of the duration of the pass-by ($L_{10}$) are given on the right-hand side. Reproduced with permission from J.-P. Clairbois.[14]

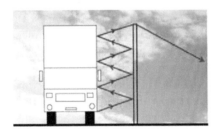

**3.15** Multiple reflections between a barrier and a high-sided vehicle

Field measurements confirming this phenomenon have been reported[13] and it has also been described by Clairbois.[14]

Figure 3.14 shows the noise level at a receiver for a single vehicle pass-by on a road without a barrier. When a reflective barrier is installed there is a reduction in noise level; on approach and departure of the vehicle, however, when the vehicle is close to the receiver there is no significant reduction in noise. On approach and departure the expected barrier attenuation is achieved as the sound travelling towards the receiver is reflected either ahead or behind the vehicle. When the vehicle is at its closest point to the receiver the reflected sound falls onto the side of the vehicle and is, in turn, reflected onto the barrier. This process continues until the sound reaches the top of the barrier where it is diffracted towards the receiver (Figure 3.15). These multiple reflections effectively raise the height of the noise source to near the top of the barrier and hence there is little or no path difference between the direct and the diffracted rays. For a steady stream of traffic, some of the sound reflected to the front or the rear of the vehicle would be reflected by other vehicles and thus the attenuation provided on approach and departure of a particular vehicle would be slightly reduced.

This phenomenon has been analysed using the boundary integral equation method.[15] It was found that the separation between the vehicle and the barrier was not critical, with the effect being present at separation distances in the range 2–22 m. The analysis was for a two-dimensional model, but the results cannot readily be translated into a three-dimensional case. The results for using a sound-absorbing face on the barrier were also reported and show that this can restore the barrier performance, even if the vehicle side is taller than the barrier.

The reduction of reflective barrier performance only significantly affects the peak noise of any vehicle pass-by and thus it affects some noise measurement indices more than others. The majority of traffic noise prediction methods are based on the equivalent continuous noise level, $L_{Aeq}$, and this is sensitive to changes in peak noise levels. The CRTN method is unique in that it uses the $L_{A10}$, (the level exceeded for 10 per cent of the analysis period) and this index is less sensitive to changes in peak noise levels. This may explain the paradox that the existence of this phenomenon has been recognised in many countries, whereas its significance has long been denied in the UK. Thus, the arbitrary choice of measurement index contributed to the use of predominantly reflective barriers in the UK until the end of the 20th century, whereas elsewhere the widespread use of absorptive barriers is commonplace. More recently this trend has been reversed and there has been a noticeable increase in the use of sound absorptive barriers in the UK.

Railways present another situation where barriers can be placed very close to the source of noise, with trackside barriers often only 1 metre from the side of the train. Morgan *et al.* used a two-dimensional numerical boundary element model to investigate the performance of railway barriers.[16] Their results suggest that, for any particular shape of barrier, absorbent barriers provide 6–10 dB better screening efficiency than the reflective equivalent.

## Improving the acoustic performance of a screen

### Introduction

There are often many conflicting design factors which have to be considered when specifying a barrier. The landscape architect may welcome some visual screening provided by a relatively small noise barrier, but could be concerned that a tall barrier could be visually intrusive to those it is designed to protect and oppressive to the driver. At the same time it must be recognised that increasing the height of a barrier can add disproportionately to the costs of the structure. Thus, there have been pressures on the acoustic engineer to provide ever more ingenious solutions to enhance the attenuation provided by barriers without increasing their height. Below is a summary of both well-established techniques and also more recent novel solutions.

### Cantilevered barriers

This is the simplest solution to the problem of reducing barrier height, whereby the top section of the barrier is angled towards the traffic. This enables the diffracting edge of the barrier to be placed considerably closer to the source of the noise than in the case of a vertical barrier. Examples range from simple cantilevers which move the top edge of the barrier only one or two metres, to large canopies which extend over one or more of the traffic lanes. This type of barrier has become increasingly popular in Asia for screening tall buildings; however, Kim *et al.*[17] used the geometrical theory of diffraction to demonstrate that the benefits afforded are not uniform throughout the height of the building. Significant increases in attenuation are achieved at lower and upper floors, but at some middle floors the effective height of the barrier is reduced compared with a vertical barrier of equivalent length and the attenuation can be reduced.

### Tunnels

A tunnel offers the most effective means of screening traffic noise. It is the cost of a tunnel which generally prohibits its use, but nevertheless tunnels are becoming a more common solution in areas of particular sensitivity. The design may range from relatively lightweight covers which provide just enough attenuation of sound to enable a noise limit to be met at nearby sensitive properties, to substantial load bearing structures, which permit the space above the tunnel to be used for a variety of purposes.

Care must be taken to ensure that the reverberant noise build up within a tunnel does not cause increased noise in the vicinity of the tunnel portals. Lining the tunnel walls and roof with sound absorptive material will avoid this problem and Woehner[18] recommended that a length of tunnel equal to two to three times the tunnel diameter should be lined at the portal.

### Galleried barriers

A galleried barrier is a hybrid solution in which a substantial cantilevered barrier completely covers the nearside traffic lanes, effectively placing them in a tunnel, and then extends over some or all of the far-side traffic lanes. This is a highly effective form of barrier and does not have the disadvantages of increased noise and decreased air quality associated with tunnel portals.

### Louvred covers

Partial tunnels can be created by suspending arrays of sound-absorbing panels above the carriageway. These panels are hung vertically, usually in diamond patterns or parallel rows, and are arranged to prevent any direct line of sight to the traffic when viewed at an oblique angle from adjacent noise sensitive properties. Thus, the direct sound rays will be absorbed, and only the rays diffracted at the edges of the panels will reach the receptors beside the road. This solution again avoids the problems of increased noise and air pollution associated with tunnel portals, and has the additional advantage of allowing natural light onto the roadway.

### Sonic crystals

In the mid 1990s, Martínez-Sala *et al.* showed that periodic arrays of cylindrical rods could be used to attenuate the transmission of sound at certain frequency ranges.[19] The attenuation mechanism was identical to that found in photonic crystals[20] and thus these minimalist structures or sculptures became known as sonic crystals. A logical development of the sculpture was to elongate the periodic array to form a barrier. Sanchez-Perez *et al.* reported the results of outdoor tests on a 7.2 metre long array, formed from 3 metre long cylindrical rods arranged in a triangular pattern, with a filling fraction of 47 per cent, and showed that two-dimensional arrays of hollow cylinders could provide similar attenuation of sound to that of conventional barriers at certain frequencies.[21] The results showed that up to 25 dB attenuation could be achieved over a wide range of frequencies between 800 Hz and 4 kHz, but this was not uniform and the attenuation was as low as 7 dB at intervening frequencies. This work was extended to investigate the performance of trees arranged in periodic lattice patterns and it was found that, particularly at frequencies below 500 Hz, the lattices performed significantly better than normal forest planting, indicating that planting patterns should also be considered when planting screen belts of trees.

The problem of the characteristic non-uniform attenuation achieved by arrays of rigid cylinders was addressed by Umnova *et al.* who studied the effects of porous coverings on the cylinders.[22] Their results showed that the porous coverings did make the attenuation provided by the array more uniform when compared with a similar array of rigid cylinders. In addition, the overall attenuation was also improved.

### Diffracting-edge modifications

The overall acoustic performance of a vertical screen is generally governed by the diffraction of sound at its top edge. It is, therefore, not surprising that considerable research has been undertaken to refine the design of this top edge to maximise the potential diffraction of the grazing sound ray. The main categories of diffracting-edge treatments are described below. It should be noted that many of the systems described are protected by patent.

#### T-shaped barriers

Several researchers have reported the effects of installing a horizontal cap on top of a vertical screen, thus forming a T-shaped barrier. This profile is effectively a small cantilevered barrier and, as such, it may be expected that there would be an improvement in performance. The reported benefits are, however, greater than would be attributable to the cantilever alone. Hothersall *et al.*[23] defined the effective height of the T-shaped barrier as the intersection of the direct ray from the source grazing the closest edge of the top and the projection of the plane of the vertical barrier. A 1 metre wide sound absorptive cap placed on a vertical screen gave about 1 dB(A) greater attenuation than a thin vertical screen equal in height to the effective height of this T-shaped barrier.

In full scale model tests, the Transport Research Laboratory (TRL) reported benefits of 2–3 dB(A) when using 1 metre and 2 metre wide tops which were faced with sound absorption, and a 1.4 dB(A) benefit for a 1 metre wide reflective top.[24] These results are broadly consistent with other researchers[25,26] and are greater than can be explained by the cantilever effect.

### Multiple-edge barriers

Many of the researchers who studied the performance of T-shaped barriers also investigated the benefits of using multiple diffracting edges. Again, both reflecting and sound absorptive systems have been extensively examined. The results of full-scale-model testing at TRL are given in Figure 3.16 and typically show about 2.5 dB(A) benefits for screens with three diffracting edges. It should be noted that these improvements are expressed as relative to a vertical screen of equal height, and no allowance is made in these results for the additional attenuation that would be expected from a vertical screen of the effective height of the multiple-edged barrier. TRL subsequently tested a multiple-edged barrier *in situ* on the M25 and Watts[27] reported benefits of up to 3 dB(A), although the additional attenuation decreased significantly for barriers which already gave high attenuations.

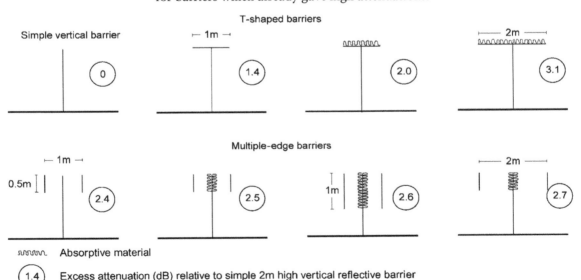

**3.16** Relative insertion losses of T-shaped and multiple-edge barriers compared with a single vertical barrier (adapted from Watts[27])

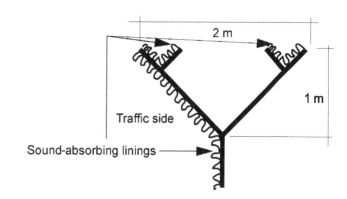

**3.17** A multiple Y-shaped barrier
(adapted from H. Shima *et al.*[28])

2 m

1 m

Traffic side

Sound-absorbing linings

## Y-shaped barriers

A novel Y-shaped barrier has been developed in Japan which has two small
Y-shapes at the ends of a larger Y-shaped barrier, thus creating four diffracting edges (Figure 3.17). The traffic side of the barrier and the insides of the
small Y-shapes are lined with sound absorption. Shima *et al.*[28] report the
results of full-scale-model tests which gave substantially better results for
this novel barrier than for either a plain Y-shape or a vertical barrier of equal
height. As much as 10 dB benefit was reported over the performance of a
vertical barrier at about 500 Hz and this is attributed to destructive interference of sound. Again, it must be noted that the results tend to overstate the
benefit as no allowance is made for the effective height of the equivalent
vertical screen.

## Tubular cappings

The use of tubular elements installed as a capping on top of vertical barriers
has been studied by many researchers. The most popular forms are either
cylindrical or mushroom-shaped and are fabricated from reflective or sound
absorptive materials.

Alfredson and Du[29] used a boundary element model and reported a
3 dB(A) benefit for an absorptive cylinder, when compared with the
performance of a reflective vertical screen of the same overall height. No
significant difference was found between the vertical screen and one capped
with a reflective cylinder. Fujiwara and Furuta[30] used theoretical, scale-
model and full-scale *in situ* tests to conclude that a 0.5 m diameter absorptive cylinder capping gave 2–3 dB(A) benefit, with the greatest effect being
measured in the range 1–1.6 kHz. They note that, although modest, the
benefit is equivalent to that which would be obtained by raising the height of
the vertical screen by some 2 m. Practical developments of the cylindrical
capping are a mushroom section, usually about 0.6 m wide, which may be
readily clamped to the top of existing vertical barriers (Figure 3.18 and an
octagonal section, Figure 3.19). The consensus of the research is that these
devices give a benefit of 2–3 dB(A).[31–33]

**3.18** Noise reducer (redrawn from Nitto Boseki Co. Ltd., Tokyo brochure *Noise Reducer Technical Information*). Picture courtesy of Tonomao Okubo at the Kobayasi Institute of Physical Research

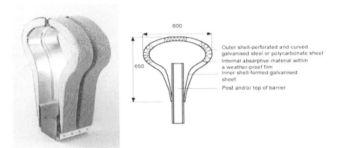

Outer shell-perforated and curved galvanised steel or polycarbonate sheet
Internal absorptive material within a weather-proof film
Inner shell-formed galvanised sheet
Post and/or top of barrier

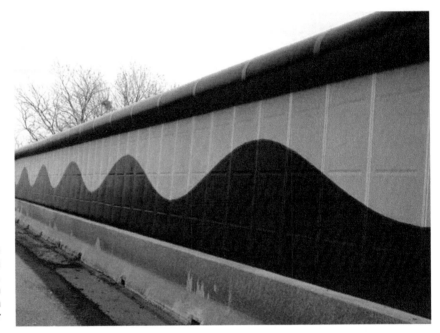

**3.19** Tubular absorptive noise reducer on top of an absorptive barrier at Ghent, Belgium. Photograph courtesy of Barbara van Hoorewerder

### Passive phase interference devices

The phenomenon of destructive interference of sound, whereby two similar sounds which are out of phase cancel each other out, is well known. A passive device that utilises this effect has been developed from work carried out by Mizuno *et al.*[34,35] and further developed by Iida *et al.*[36] (Figure 3.20). The device comprises an open-topped box section, 0.5 m wide and 0.7 m deep, which is fitted at the top rear part of a vertical barrier. The sound diffracted at the top edge of the barrier enters the box via four angled channels and, because of the uneven lengths of these channels, destructive interference occurs at the outlets. The device is claimed to give up to 6 dB(A) increases in attenuation, but Watts carried out full-scale-model and *in situ* tests and reported somewhat smaller benefits.[37] Watts found that the device gave an average benefit of 1.9 dB(A), but of this he attributed only 0.7 dB(A) to the interference effect, with the remainder being due to diffraction at the front and rear edges of the device.

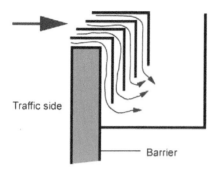

**3.20** Phase interference barrier capping (adapted from Iida et al.[36])

Traffic side

Barrier

An alternative approach was investigated by Fujiwara and Yotsumoto[38] in which they covered a reflective cylindrical barrier cap with an array of tubes. The length of the tubes was equal to a quarter of the wavelength of the sound to be cancelled, and destructive interference occurred at the outlets of the tubes. This interference takes place at a discrete frequency, and a practical barrier would require a range of tubes of varying length.

Fujiwara et al.[39] have described the results of boundary element modelling of a T-shaped barrier with a reactive surface comprising an array of tubes or wells on the upper surface. The well depths were tuned to 420 Hz for one model and two depths corresponding to 420 Hz and 840 Hz were used in the second model (Figure 3.21). Substantial benefits over a simple T-shaped barrier of 5 dB at low frequencies and 10 dB or more at frequencies above 1 kHz are claimed. The results of two-dimensional scale-model tests are also reported and there is good agreement between the results. Full-scale-model tests on this type of barrier have not been reported, but it may be expected that the results of such tests would be more modest.

A more complex barrier, that incorporated one or more sets of waveguide filters, was described be Amram et al.[40] The low-pass filters act as a dipole line source which effectively retards the transmitted ray, and destructive interference subsequently occurs with the ray diffracted over the barrier (Figure 3.22). Scale- and full-size-model tests were used and, when

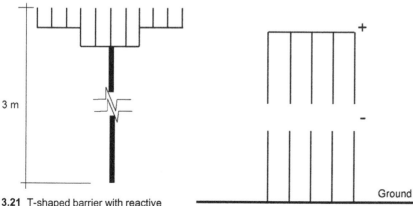

3 m

**3.21** T-shaped barrier with reactive surface (adapted from Fujiwara et al.[39])

+

−

Ground

**3.22** Basic dipolar-type barrier

compared with a solid vertical barrier of similar height, excess attenuations of up to 5 dB were reported at low frequencies; however, not surprisingly, the performance was degraded at some higher frequencies. The overall performance of the barrier was not reported, but the frequency spectra given suggest that this was minor and further development work does not appear to have taken place.

Active noise control

The excess attenuation that has been achieved by use of the diffracting-edge modifications described above is relatively modest, and potentially far greater attenuations may be realised using active noise control. This technology has now been successfully applied in several industries, especially where the propagating sound field is a plane-wave form. An array of microphones is used to monitor the sound field and feed an array of loudspeakers, located between the source and the receiver, with a similar sound field that is 180° out of phase with the source. This has been studied by Krutzen *et al.*[41] and de Beer *et al.*[42] The studies were conducted as part of the Dutch Noise Innovation Programme with the aim of producing a practical active noise barrier that would provide a reduction of 5 dB(A) at up to 100 m behind the barrier. The attenuation at a range of receivers was investigated using a line of loudspeakers located along the top edge of the community side of the barrier diffracting-edge, with an array of sensor microphones along the road side of the barrier. The study showed that the target attenuation could be achieved at up to 75 m behind the barrier, with a significantly greater performance of up to 13 dB(A) at 25 m. Two heights of barrier were used, 3 and 6 m, and the greatest enhancements in performance were found with the taller barrier. Another significant result was that the performance was greatest at lower frequencies, below 1 kHz, which resulted in greater reductions in dB(A) noise level being achieved within dwellings than outside, due to the high-frequency bias of the attenuation of building elements. A hybrid barrier modification has been developed in Japan by Mitsubishi Heavy Industries Ltd that combines a sound absorptive passive element to enhance high-frequency attenuation with active noise control, provided by a line array of loudspeakers to increase low-frequency attenuation (Figures 3.23 and 3.24).

**Noise leakage through gaps**

Where a barrier is constructed from more than one type of material, the overall insertion loss of a barrier will depend upon the sound transmission loss of each element and the ratio of the areas of the different elements. If large holes are present, their transmission loss is zero and could seriously compromise the performance of the barrier. It is more likely that narrow slits may be present in a barrier, due to poor sealing between the panels and the support posts, or between the panels themselves. Early theoretical work by Gomperts and Kihlman[43] revealed that narrow slits act as preferential frequency filters: for most frequencies the transmission loss is positive but at certain frequencies there may be a significant negative transmission loss. The

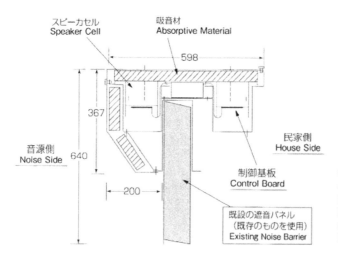

**3.23** Cross-section of the Active Soft-edge Barrier by Mitsubishi Heavy Industries Ltd., Japan. Information courtesy of Mitsubishi Heavy Industries Ltd., Japan

**3.24** Installed Active Soft-edge Barrier by Mitsubishi Heavy Industries Ltd., Japan. Photograph courtesy of Mitsubishi Heavy Industries Ltd., Japan.

negative transmission loss is due to resonances which occur where the depth of the slit is equal to half of the wavelength of the incident sound. Later work by Mechel[44] showed that filling the slit with a sound-absorbing material would damp out the resonances and ensure that the transmission loss was positive at all frequencies.

The reduction in performance of a timber barrier caused by the presence of gaps between the panels was investigated by Watts.[45] Using typical gap sizes found in timber barriers it was found that for barriers up to 3 m high the degradation in performance was less that 1 dB(A) at distances over 20 m. More significant reductions in performance were found with taller barriers (up to 6 m), where reductions of more than 2 dB occurred at over 20 m.

Thus, for most practical situations, the presence of a narrow slit will not significantly reduce the overall insertion loss of the barrier, since the area of the opening will be a very small percentage of the total area, but more care must be taken to avoid gaps as taller barriers become increasingly common. The sound field close to a barrier containing empty slits will, however, sound distorted due to the enhanced transmission of discrete frequencies, and therefore joints should be made using a compressible seal.

There are occasions when it may be desirable to leave a narrow gap in a barrier because of an engineering or environmental constraint. In Finland, for example, research has been carried out on the effects of leaving a gap of up to 0.4 m at the bottom of a barrier to avoid damage from ground heave during frosts. The results show that, although the insertion loss of the barrier was reduced by up to 3 dB immediately behind the barrier, the performance at 20 m was not significantly compromised.

Gaps may also be left in barriers for emergency access and maintenance. Where this is done the acoustic performance of the barrier should be preserved by erecting a parallel, overlapping barrier behind the gap. One face of the overlap must be lined with sound-absorbing material to prevent reverberation within the gap. The length of the overlapping section should be at least as long as the horizontal separation between the two barriers, although research in California suggests that the overlap should be 2.4 times the separation[46] (Figure 3.25).

**Varying longitudinal profiles**

The use of a saw-tooth longitudinal profile barrier offers the possibility of destructive interference taking place between the sound transmitted between the teeth and the sound diffracted over the top of the barrier. Wirt[47] carried out scale-model tests using saw teeth or thnadners on top of a solid wall and reported benefits of 1.5–4.5 dB(A) when compared with a wall equal in height to the complete barrier. May and Osman[12] and Hutchins et al.[1,2] studied the use of thnadners alone and reported poorer results than for the equivalent solid barrier. Their work did reveal that the loss in performance was least when the thnadners were faced with sound absorption.

Ho et al.[48] have carried out scale-model tests on randomly jagged-edged barriers and reported that these give enhanced high-frequency performance

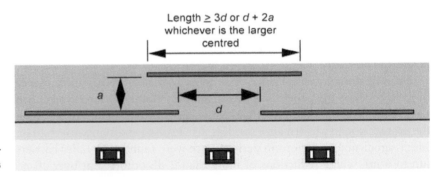

Length ≥ 3d or d + 2a
whichever is the larger
centred

a

d

3.25  Typical arrangement for overlapping parallel barriers

but poorer low-frequency performance when compared with solid straight-edged barriers of equal overall height. The degree of excess attenuation was found to be directly proportional to the degree of jaggedness and an empirical formula was derived to describe the phenomenon. Further work has been carried out by Shao et al.,[49] on developing a theoretical model for the performance of random-edge profiled barriers. The performance of the barrier was found to depend on the average height and the variation in the randomness of the edge. Their theoretical results, which were validated by scale-model tests, showed that in certain cases a random-edge barrier could provide greater attenuation than a straight-edged barrier with a height equal to the average height of the random-edge barrier.

### Noise screening by vegetation

The use of vegetation to provide screening has been studied by many researchers. This is generally restricted to field measurement studies, although some tentative attempts have been made to develop mathematical models to describe the phenomenon.

Kragh[50] reported the results of field studies on seven belts of vegetation, which varied between 15 m and 40 m in depth, and comprised mainly deciduous trees and bushes. The excess attenuation provided was determined by measuring in front of and behind the tree belt and comparing the difference with that calculated for grassland. The benefits were greatest at low frequencies (around 250 Hz) and at high frequencies (>1 kHz). The excess attenuation at these frequencies was often 6–8 dB but little or no benefit was recorded at mid-frequencies, which are the dominant component of traffic noise, and therefore the overall excess attenuation was typically 3 dB$L_{Aeq}$ or less. Broadly similar results were reported by Huddart[51] who found that a 10 m deep belt of dense spruce gave an excess attenuation of 6 dB$L_{A10}$. This apparent discrepancy is partly explained by the use by the authors of different noise indices. Huddart showed that the highest excess attenuation occurred at the upper percentiles and that there was typically 1–2 dB more in the $L_{A10}$ index than in the $L_{Aeq}$.

The improvement in low-frequency attenuation is due to the enhanced ground absorption caused by fallen leaves and branches increasing the surface porosity. The vegetation itself is considered to be responsible for the high-frequency attenuation. Martens[52] observed that the maximum attenuation occurred where the leaf dimension was equal to half the wavelength of the sound, and consequently it is essential that any vegetative screen contains a high proportion of deciduous species.

There are some obvious drawbacks to using vegetative screens as the sole means of noise mitigation. Perhaps the main disadvantage if used for screening property is that the benefit is not immediate and it may take many years for the planting to develop, and in the meantime it may be necessary to provide other mitigation, such as noise insulation or a fence which may be allowed to deteriorate as the vegetation grows. By its very nature a tree belt is a constantly changing screen and in time may become more or less effective, and it is, therefore, not possible to precisely define the attenuation which will be provided. Thus, tree belts are not currently considered where accurate calculations have to be made for noise insulation assessments. A further consideration is that a degree of maintenance is required throughout the life of the screen if its optimum performance is to be achieved. There is evidence that the presence of vegetation on an earth mound provides a more effective screen than either the mound or the vegetation alone.[53] Consequently the role of vegetation is likely to be either to enhance the performance of earthworks or to quieten open countryside where a precise reduction in noise is not required.

## Acoustic performance testing

The European Committee for Standardization (CEN) has published a standard, EN 1793, which defines test procedures for measuring the acoustic performance of noise barriers and this has been adopted as a British Standard.[54-56] Part 1 of this standard gives a laboratory test method for measuring the sound absorption of a barrier, and Part 2 gives a laboratory test method for measuring the airborne sound insulation or transmission loss of a barrier. Both tests allow a single-number rating to be derived for the barrier and to calculate these it is necessary to use the standardised traffic noise frequency spectrum, which is given in Part 3 of the standard.

Laboratory testing does allow measurements to be made under controlled conditions, but there are certain drawbacks. The test sample is invariably perfectly installed in the test chamber in order to yield the optimum result, but this may not be representative of the performance that will be achieved by a real installation. The test sample may be stiffer than the installed barrier as the panel elements will generally need to be cut to fit the test aperture, and also the top edge will be constrained within the aperture. The sound absorption test follows the procedure currently in use for the testing of other acoustic materials and allows the sample to be laid on the floor of the test chamber. This is acceptable for rigid, homogenous materials, but can yield misleading results for composite barrier systems.

To reduce the uncertainties introduced by laboratory testing, the CEN working group has developed two *in situ* test procedures for barriers. The first allows both the sound absorption and the airborne sound insulation of barriers to be tested on site, and also provides a means of checking that the desired acoustic performance of a barrier is maintained throughout its life.[57] Furthermore, it will allow the acoustic performance of bio-barriers to be quantified, since it is not practical to test these in a laboratory.

The second document describes a test method for determining the intrinsic characteristics of sound diffraction of added devices installed on top of traffic-noise-reducing devices.[58] This standard test procedure is particularly welcome since many highway authorities were reluctant to specify any of the growing range of these devices in the absence of an agreed method for quantifying their acoustic performance. The test method prescribes measurements of the sound pressure level at several reference points near the top edge of a noise-reducing device with and without the added device installed on top of it. Impulse response measurements are made to extract only the direct diffraction from the edge and not the sound reflected from the ground or any other reflecting surface. The effectiveness of the added device is calculated as the difference between the measured values with and without the added devices, correcting for any change in height. The CEN/TS methodology allows the intrinsic efficiency of top-edge modifications to be evaluated and compared; however, the single figure ratings of device efficiency cannot be accurately combined with ground conditions to evaluate complete barrier performance in particular locations. Okubo and Yamamoto have described a theoretical model for evaluating the sound field behind an edge-modified barrier.[59] Their method is based on the sum of four diffraction paths, including the ground reflection, and the intrinsic efficiency of the edge-modification device is substituted into each path.

## References

1. Hutchins, D.A., Jones, H.W. and Russell, L.T. (1984) 'Model studies of barrier performance in the presence of ground surfaces, Part 1–Thin perfectly reflecting barriers'. *Journal of the Acoustical Society of America*, 75(6), 1807–16.

2. Hutchins, D.A., Jones, H.W. and Russell, L.T. (1984) 'Model studies of barrier performance in the presence of ground surfaces, Part 2–Different shapes'. *Journal of the Acoustical Society of America*, 75 (6), 1817–26.

3. Maekawa, Z. (1968) 'Noise reduction by screens', *Journal of Applied Acoustics*, 1, 157–73.

4. Rathe, E.J. (1969) 'Note on two common problems of sound attenuation'. *Journal of Sound and Vibration*, 10(3), 472–9.

5. Kurze, U.J. and Anderson, G.S. (1971) 'Sound attenuation by barriers', *Applied Acoustics*, 4, 35–53.

6. Department of Transport and Welsh Office (1988) *Calculation of Road Traffic Noise*, HMSO, London.

7. Department of Transport (1976) *Noise Barriers – Standards and Materials Technical Memorandum H14/76*.

8. Hothersall, D.C., Chandler-Wilde, S.N. and Crombie, D.H. (1993) 'Modelling the performance of road traffic barriers', Proc TRL meeting, *Traffic Noise Barriers*, pp. 22–37.

9. Watts, G.R. (1995) 'Acoustical performance of parallel Traffic Noise Barriers', *Applied Acoustics*, 47 95–119.

10. Tobutt, D.C. and Nelson, P.M. (1990) *A Model to Calculate Traffic Noise Levels from Complex Highway Cross-sections*. Report RR 245, Transport and Road Research Laboratory, Crowthorne.

11. Slutsky, S. and Bertoni, H.L. (1988) 'Analysis and programmes for assessment of absorptive and tilted parallel barriers', *Transport Research Record*, Washington DC.

12. May, D. & Osman, M. (1980) 'Highway noise barriers: new shapes', *Journal of Sound and Vibration*, 71(1), 73–101.

13. Yamashita, M., Kaku, J. and Yamamoto, K. (1985) 'Net effects of absorptive acoustic barrier', *Inter Noise*, Munich pp. 507–10.

14. Clairbois, J-P. (1990) *Road and Rail Noise–Corrective Devices*. Seminar on Acoustic Noise Barriers–The Engineered Solution to Road and Rail Noise Pollution. I Mech E, London.

15. Hothersall, D.C. and Tomlinson, S.A. (1995) 'High sided vehicles and road traffic noise barriers', *Proceedings of Internoise 95*, Newport Beach, USA, pp. 397–400.

16. Morgan P.A., Hothersall D.C and Chandler-Wilde S.N. (1998) 'Influence of shape and absorbing surface – a numerical study of railway noise barriers', *Journal of Sound and Vibration*, **217**, 405–17.

17. Kim H-S., Kim J-S., Kang H-J., Kim B-K. and Kim S-R. *Applied Acoustics*, **66**(9), September 2005, pp. 1102–119.

18. Woehner, H. (1992) 'Sound propagation at tunnel openings', *Noise Control Engineering Journal*, **39** (2), 47–56.

19. Martínez-Sala M., Sancho J., Sánchez J.V., Gómez V., Llinares J. and Meseguer F. (1995) 'Sound attenuation by sculpture', *Nature*, **378**, 241.

20. 'Photonic Crystals are wavelength scale, periodic, dielectric microstructures. Their periodic patterning creates photonic band gaps which forbid the propagation of light through the structure.' http://www.st-andrews.ac.uk/~photocryst/.

21. Sanchez-Perez, J.V., Rubio, C., Martinez-Sala, R., Sanchez-Grandia, R. and Gomez, V. (2002) 'Acoustic barriers based on periodic arrays of scatterers', *Applied Acoustics Letters*, **81**, 240–2.

22. Umnova, O., Attenborough, K. and Linton, C.M. (2006) 'Effects of porous covering on sound attenuation by periodic arrays of cylinders', *Journal of the Acoustical Society of America*, **119**(1) 278–84.

23. Hothersall, D.C., Crombie, D.H. and Chandler-Wilde, S.N. (1991) 'The performance of T-profile and assorted noise barriers', *Applied Acoustics*, **32**, 269–87.

24. Watts, G. (1993). 'Acoustic performance of new designs of traffic noise barriers', *Proc Noise '93*, St Petersburg, Russia.

25. Hajeck, J.J. and Blaney, C.T. (1984) 'Evaluation of T-profile highway noise barriers', *Transport Research Record*, Washington DC.

26. May, D.N. and Osman, M.M. (1980) 'The performance of sound absorptive, reflective and T-profile noise barriers in Toronto', *Journal of Sound and Vibration*, **71**(1), 65–71.

27. Watts, G.R. (1996) 'Acoustic performance of a multiple-edge noise barrier profile at motorway sites', *Applied Acoustics*, **42**, 47–66.

28. Shima, H. Watanabe, T. Mizuno, K. Iida, K. Matsumoto, K. and Nakasaki, K. (1996) 'Noise reduction of a multiple-edge noise barrier', *Proceedings of Internoise 96*, Liverpool, pp. 791–4.

29. Alfredson, R.J. and Du, X. (1995) 'Special shapes and treatment for noise barriers', *Proceedings of Internoise 95*, Newport Beach, CA, pp. 381–4.

30. Fujiwara, K. and Furuta, N. (1991) 'Sound shielding efficiency of a barrier with a cylinder at the edge', *Noise Control Engineering Journal*, **37**(1) 5–11.

31. Fujiwara, K. Ohkubu, T. and Omoto, A. (1995) 'A note on the noise shielding efficiency of a barrier with absorbing obstacle at the edge', *Proceedings of Internoise 95*, Newport Beach, CA, pp. 393–6.

32. Yamamoto, K., Shono, Y., Ochiai, H. and Yoshihiro, H. (1995) 'Measurements of noise reduction by absorptive devices mounted at the top of highway noise barriers', *Proceedings of Internoise 95*, Newport Beach, CA, pp. 389–92.

33. Gharabegian, A. (1995) 'Improving soundwall performance using route silent', *Proceedings of Internoise 95*, Newport Beach, CA, pp. 385–8.

34. Mizuno, K., Sekiguchi, H. and Iida, K. (1984) 'Research on a noise control device – first report, fundamental principles of the device', *Japanese Society of Mechanical Engineers*, 27(229) 1499–505.

35. Mizuno, K., Sekiguchi, H. and Iida, K. (1985) 'Research on a noise control device – second report, fundamental design of the device', *Japanese Society of Mechanical Engineers*, 28(245), 2737–43.

36. Iida, K., Kondoh, Y. and Okado, Y. (1984) 'Research on a device for reducing noise.' *Transport Research Record*, 983, 51–4.

37. Watts, G. (1996) 'Acoustic performance of an interference-type noise-barrier profile', *Applied Acoustics*, 49(1) 1–16.

38. Fujiwara, K. and Yotsumoto, E. (1990) 'Sound shielding efficiency of a barrier with soft surface', *Proceedings of Internoise 90*, pp. 343–6.

39. Fujiwara, K. Hothersall, D.C. and Kim, C-H. (1996) 'Noise barriers with reactive surfaces', *Applied Acoustics*, 53(4) 255–72.

40. Amram, M., Chvrojka, V.J. and Droin, L. (1987) 'Phase reversal barriers for better noise control at low frequencies: laboratory versus field measurements', *Noise Control Engineering Journal*, 28(1), 16–23.

41. Krutzen, M.P.M., Mast A. and de Beer, F.G. (2002) *Active Noise Barrier: Phase1: Simulation Study*, TNO Report DGT-RPT-020050, TNO, Delft, the Netherlands.

42. de Beer, F.G., Berkhoff, A.P., Golliard, G., van Lier, L.J., Salomons, E.M., van der Torn, J.D., Vedy, E.M.P., and Vermeulen, R.C.N. (2004) *Active Noise Barrier: Extended Simulation Study*, TNO Report DGT-RPT-040025, TNO, Delft, the Netherlands.

43. Gomperts, M.C., and Kihlman, T. (1968) 'The sound transmission loss of circular and slit-shaped apertures in walls', *Acustica*, 18, 144–50.

44. Mechel, F.P. (1986) 'The acoustic sealing of holes and slits in walls', *Journal of Sound and Vibration*, 111(2), 297–336.

45. Watts, G.R. (1999) *Effects of Sound Leakage through Noise Barriers on Screening Performance*, Sixth International Congress on Sound and Vibration, Copenhagen, Denmark.

46. The Transportation Research Board (1982) *Highway Noise Barriers*, National Academy of Sciences, Washington, D.C., p. 15.

47. Wirt, L.S. (1979) 'The control of diffracted sound by means of thnadners (shaped noise barriers)', *Acustica*, 42(2) 73–88.

48. Ho, S.S.T., Busch-Vishniac, I.J. and Blackstock, D.T. (1997) 'Noise reduction by a barrier having a random edge profile', *Journal of the Acoustical Society of America*, 101(5) pt 1, pp. 2669–76.

49. Shao W., Lee H.P., and Lim S.P. (2001) 'Performance of noise barriers with random edge profiles', *Applied Acoustics*, 62, 1157–70.

50. Kragh, J. (1982) *Road Traffic Noise Attenuation by Belts of Trees and Bushes*. Danish Acoustical Laboratory Report no. 31.

51. Huddart, L. (1990) *The Use of Vegetation for Traffic Noise Screening*, Transport and Road Research Laboratory Report no. 238.

52. Martens, M.J.M. (1980) 'Foliage as a low pass filter: experiments with model forests in an anechoic chamber', in Martens, M.J.M. (ed). *Geluid en Groen*. Katholieke Universiteit, Nijmegen, Netherlands, Ch 6, pp. 118–40.

53. Cook, D.I. and Van Haverbeke, D.F. (1974) *Tree Covered Landforms for Noise Control*. The Forest Service, US Department of Agriculture. Research Bulletin 263.

54. European Committee for Standardization (1997) *EN 1793–1 Road Traffic Reducing Devices. Test Method for Determining the Acoustic Performance. Part 1: Intrinsic Characteristics of Sound Absorption*, CEN, Brussels.

55. European Committee for Standardization (1997) *EN 1793–1 Road Traffic Reducing Devices. Test Method for Determining the Acoustic Performance. Part 2: Intrinsic Characteristics of Airborne Sound Insulation*, CEN, Brussels.

56. European Committee for Standardization (1998) *EN 1793–3 Road Traffic Reducing Devices. Test Method for Determining the Acoustic Performance. Part 3: Normalized Traffic Noise Spectrum*, CEN, Brussels.

57. European Committee for Standardization (2003) *CEN/TS 1793–5 Road Traffic Noise Reducing Devices. Test Method for Determining the Acoustic Performance. Part 5: Intrinsic Characteristics. In Situ Values of Sound Reflection and Airborne Sound Insulation*, CEN, Brussels.

58. European Committee for Standardization (2003) *CEN/TS 1793–4 Road Traffic Noise Reducing Devices. Test Method for Determining the Acoustic Performance. Part 4: Intrinsic Characteristics. In Situ Values of Sound Diffraction*, CEN, Brussels.

59. Okubo T. and Yamamoto K. (2007) 'Simple prediction method for sound propagation behind edge-modified barriers', *Acoustical Science and Technology* 28, 1.

# Barrier morphology and design

4

## Anatomy of barrier – elements and characteristics

It has been necessary to create a simple morphology for barriers in order to describe their component parts and analyse the elements that help to make or break their aesthetic appeal. With this tool it is possible to describe barriers, directing attention towards their individual elements and clarifying why some barriers are more visually appealing or satisfactory than others.

Essentially, barrier morphology, or the classification of the forms and structures relating to barriers, is uncomplicated, as barriers are generally made up of a small number of essential parts. A barrier may be seen to have a top section, a middle section and a base section, even in cases where it is constructed out of a single uniform material. This principle is confirmed and used in the development of the 'modular' barrier system in the Netherlands (Figures 4.1 and 4.2. Refer also to Chapter 2). A barrier also has a top edge that provides a silhouette against its background, a bottom edge where it meets the surrounding ground, a support structure and foundations (Figure 4.2).

### Top section and top edge

The top section and edge of the barrier is critical to its visual appeal and appearance when seen against the landscape/townscape backdrop. For a pedestrian, with an average eye level of approximately 1.5–1.7 m, the top of a barrier is nearly always going to be above eye level. This means that, more often than not, the top edge of the barrier will be silhouetted against the sky, a backdrop of vegetation or perhaps against buildings and other built forms. In a car, eye level is considerably lower, at approximately 1.3–1.4 m, and thus barriers are always likely to be viewed upwards. The background against which the barrier will be viewed is, therefore, an important consideration and it is essential to ascertain whether the visual and aesthetic strategy calls for losing the top edge of the barrier within the landscape or alternatively providing some strength to the edge so that it makes a more dominant visual statement (Figures 4.3 and 4.4).

1.

2.

3.                                                          4.

**4.1** A range of 'modular barriers' as part of air pollution studies in the Netherlands: 1. Working face of a profiled woodfibre concrete barrier with four modules; 2. Community side façade of a semitransparent barrier with four modules. 3. Roadside façade of the same semitransparent barrier. 4. Roadside face of a painted steel sheet barrier

Usually the answer depends upon the location of the barrier. Generally there are two main scenarios: countryside and cityscape. In the countryside or in rural areas, experience has indicated that barriers should be concealed, or appear transparent and as lightweight as possible in the landscape (Figure 4.5). The top section and top edge should fade or blend easily into the backdrop of the sky or vegetation. This may also be the case in urban areas where many successful barriers use transparent and lighter materials at the top to reduce the overall apparent height of the barrier and to allow light to pass through (Figures 4.6 and 4.7).

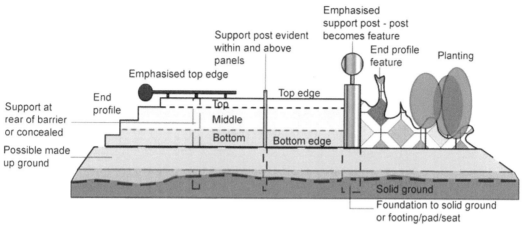

**4.2** Barrier morphology

**4.3** Non-acoustic element forming a top edge

**4.4** The visual image of a barrier is the sum of the barrier and its background

However, in urban areas, where there is often a jumble of built forms and a clutter of urban paraphernalia, such as signage, poles and lighting, with a juxtaposition of often discordant materials, a barrier with strength of form and a firm and distinctive silhouette may be better suited. In these situations, a well-designed barrier may help to strengthen the urban structure, whereas a visually weak one may have the reverse effect and add to visual clutter and discordance (Figure 4.8).

The treatment of the top edge may, however, be dictated by acoustic considerations rather than aesthetic ones, since the diffraction of noise at the top edge may be better controlled by a number of proprietary devices. These systems generally involve the use of sound-absorbing elements such as the bulbous, mushroom form developed in Japan (Figures 3.16–3.21) or additional diffracting edges. All of these devices add to the bulk at the top edge of the barrier and require particular attention to ensure that they are visually acceptable (Figure 4.3). It is interesting to note that most current barriers do not have

**4.5** A simple transparent barrier at a bus stop in a semirural area appears as inconsequential as possible

**4.6** Arrays of transparent barriers in an urban area limits negative visual impact

diffracting top-edge modifications; instead, the top edges are mostly designed to be as visually unobtrusive as possible. This may be due to the small 2–3 dB diminution in noise afforded by the adapted top edge, as well as to the fact that the appearance of most barriers is improved where the top section is essentially

**4.7** Transparent and light-coloured materials reduce the visual impact of a barrier

**4.8** Assertive design elements can enhance the cityscape, Docklands Light Railway, London

transparent. Transparency allows the top of the barrier to blend in with its surroundings, whether natural (sky and vegetation) or man-made (buildings and other structures) or a combination of both. Despite the top section of many barriers being transparent, and hence visually more low-key, they are not always simply-ended and some variation in profile and complexity can be created (Figures 4.9–4.11). Opaque barriers usually have simple top edges as they are visually more commanding and do not need articulation.

Due to their mass, very large barriers up to 20 m tall are usually transparent or partially so, and they often utilise more complex supporting structures, which in some cases are located at the top of the barrier. An example

**4.9** Articulation of the top of a transparent barrier using the horizontal and vertical structures

**4.10** Complex arrangement of transparent panels at the top of a large barrier

of this type protects residents from the noise from the E59 motorway in Vienna. It is hung between two end façades of adjacent buildings (Figure 4.12). The 20 metre tall screen utilises a lattice, steel truss to support the barrier from the top. A second large barrier in the same vicinity uses a

**4.11** A simple bend in the acrylic sheet barrier adds some visual complexity and dynamism to the barrier

contemporary planar glazing system where the glass screen is supported with steel arms and stainless steel fixings with tension loops at the top of the barrier (Figure 4.13).

In most cases barrier design in the first decade of the 21st century has downplayed the top edge. This trend forms part of an overall fashion to design barriers whose presence is formed by a controlled and simple sense of design, without additional frills. Visual integrity and potency relies on the overall shape, form, structure and angles of the barrier, rather than being dependent on non-functional additions. Add-ons appear out of place in the contemporary environment, and not in keeping with the physical and visual integrity of a 'form follows function' ideal.

### Middle section

The middle section, or body, of the barrier is likely to form the major visually apparent part of the barrier, since it will usually be the largest section and will probably mitigate most of the views. This is also because the principles of proportion in design suggest that the barrier should not appear top or bottom heavy. The whole of the barrier needs to be considered, including the rise and fall in the topography and the incorporation of earth embankments and mounds. In many cases, a barrier may comprise a single uniform façade of a single material, but what is below the barrier, namely the ground or the

**4.12** A 20 metre tall transparent barrier with decoration supported by a steel lattice structure hung between two buildings

**4.13** Planar glazed barrier with top section stabilised by anti-torsion arms

earth, is likely to form part of the complete barrier. Thus, many barriers in rural or semirural locations comprise an arrangement of earth mounding with a vertical section on top. The visual complexity of this arrangement may be further complicated by the addition of other materials, say a transparent section on top, which will lighten the appearance of the barrier and

potentially reduce its visual impact. Ideally, to avoid visual clutter and
disharmony, the middle section of the barrier should only comprise a single
material, although this may, of course, be subject to changes in form and/or
colour (Figure 4.14).

### Bottom section and bottom edge

The bottom edge of a barrier may in reality be obscured by planting, grasses or
perhaps by a kerb or safety fence. In the past it has been seen to be essential that
this section is designed and constructed without gaps or holes through which
noise can pass. However, there are some cases in the Netherlands where gaps
are being left at the bottom of barriers so that no special drainage is required
and so that smaller animals may pass through. Maintenance costs are also
reduced as there is no contact between the soil and the barrier and therefore
there is less corrosion.

The base also needs to be in an appropriate proportion to the sections of
barrier above. Many barriers utilise planters within the bottom section. This
helps to give weight and visual complexity to the barrier without it being
overrefined. Furthermore, the use of planters removes the planting mixture
away from contaminants such as salt used to de-ice roads in winter and from
heavy metal pollution from exhausts which tend to gravitate downwards to
the road or kerb level (Figure 4.15). As mentioned above, this section is also
where water may collect and may contain areas for road drainage. Thus, the
designer needs to be aware of potential maintenance operations within the

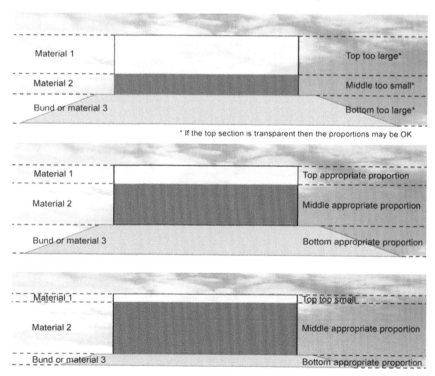

**4.14** Barrier proportions

transport corridor and of the barrier itself. The appropriate measures must be incorporated within the design to accommodate these issues.

### Barrier façades

A barrier also has two façades: the front or side that faces the traffic and the back or public side that faces the protected area. In some cases these may be treated similarly, but in most cases the front elevation differs from the rear elevation since the barrier usually mitigates noise from only one direction. The front of a barrier may also be different from the back because of the need for supports, which in many instances are placed to the rear. This difference may also be affected by a combination of other construction techniques and cost. Most importantly, however, the visual quality and character of the front and back need to reflect the visual quality and character of the particular surroundings which may differ on either side. One side of the barrier may not be as important as the other, where, for example, the barrier itself may be screened by vegetation, or where it is hidden from view. Figures 4.16 and 4.17 show the front and rear elevations of a barrier near Milan in Italy. Whereas the front façade is faced with a perforated aluminium skin, timber has been applied to the rear to soften the appearance of the barrier for the people living next to it. This effect could be further improved by planting.

An example of where the road side and community side of the barrier resemble each other is located alongside the railway line at Vleuten in the Netherlands, where one of the main priorities of the barrier design is to deter graffiti (Figures 4.18 and 4.19). The barrier has an inner core of white opaque and translucent panels. Arrays of robust galvanised steel mesh shield the inner core. Due to its open structure the mesh inhibits the spraying of graffiti. The angled noise-reducing panels (where graffiti would readily be legible) are

**4.15** Raised planters provide horticultural and visual benefits

**4.16** Road side of barrier visually reflects the motorway corridor. Note the New Jersey 'Laghi' barrier with resonators below

**4.17** Timber applied to rear side of barrier to soften the appearance for residents

**4.18** Robust measures taken to create a graffiti-free barrier at Vleuten, the Netherlands by using angled metal screens surrounding the inner core

located behind the deeply profiled mesh (where legibility of any words, tags or images is diminished due to the complexity of the steel mesh's open form).

## Ends

A barrier needs to be at its total designed height along all of its length to achieve its acoustic and visual screening objectives. Any end detail, such as tapering, must extend beyond the end point of the functional design of the barrier. Thus, whereas the design of the main façade of a barrier often concerns a dialectic between the landscape architect and the acoustic engineer over different requirements and ideals, the design of the barrier termination is

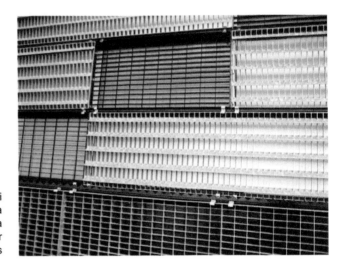

**4.19** Prototype for the anti-graffiti barrier at Vleuten, illustrating a number of mesh systems and a variety of translucent and inner core panels

**4.20** Three additional end panels are used to visually terminate the barrier. Note the use of photovoltaic panels on top

essentially an aesthetic rather than an acoustic consideration. It is usually an add-on to the full length and height required for the main barrier (Figure 4.20).

The visual transition between a route corridor without a barrier to a corridor with a barrier and *vice versa*, that is, between an open space and a visually contained space, may be awkward and therefore needs to be treated sensitively. There are two opposite ways of dealing with this situation, namely to highlight these ends or to downplay or disguise the ends. The simplest way of disguising the beginning and end of a barrier is to envelope it in planting, but it would be visually more pleasing if this planting related to other planting along the length of the barrier as well. This other planting could be located where there is an awkward height transition or a deviation in alignment (Figures 4.21).

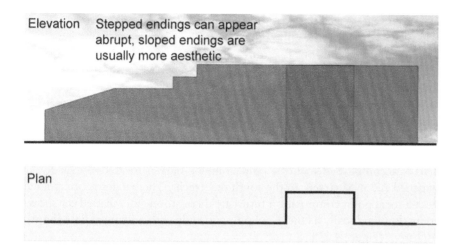

Elevation   Stepped endings can appear abrupt, sloped endings are usually more aesthetic

Plan

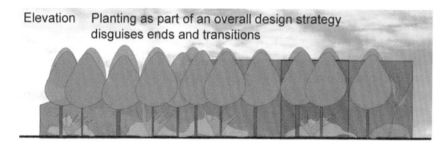

Elevation   Planting as part of an overall design strategy disguises ends and transitions

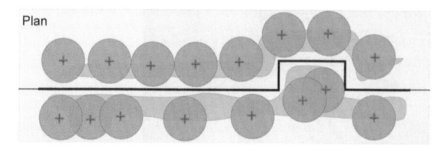

Plan

**4.21** Barrier ends and transitions

**4.22** A visual highlight to the end of a visually sophisticated semitransparent glass barrier

In most cases, well-designed barriers in rural and semirural areas do not appear to stop abruptly, although some examples show that this may be a visually effective and robust feature (Figures 4.23 and 4.24). Most barriers tend to taper, step down or break up when they start or finish. In more urban locations the ending of a barrier is often more abrupt and visually acceptable. An appropriate appearance may be achieved by the use of a feature, which integrates well with the overall barrier design (Figures 4.25 and 4.26). Generally, the visual statement that is designed for the barrier end needs to respond to the type and scale of the visual character of the barrier as a whole. This is accomplished with the large complex barrier located alongside the widened A2 at Maarsen, to the south of Utrecht. The barrier is punctuated with a focal point comprising a futuristic three-storey pod-shaped car showroom, known locally as the 'Cockpit', and ends with a flying-saucer-shaped 'full stop' (Figure 4.27).

### Vertical profile – angled, contoured and cantilevered barriers

The vertical profile of a barrier is an essential consideration. If the barrier comprises an earth mound, it is the angle of repose of the earth that forms its profile. If the barrier is made of rigid components, then these may be erected either vertically, or at an angle, or in a partially concave or cantilevered shape.

**4.23** The visual strength of the barrier is not diminished by stepping down. The transition between woodland and a stone gabion wall is visually pleasing

**4.24** An abrupt ending adds a sense of contemporaneity to this high-tech aluminium and transparent barrier

**4.25** Successfully terminated timber absorptive barrier using a sloping end

The profile of an earth mound can give the impression of a highly engineered situation if it is steep, has crisp lines and is characterised by visually simple planting, such as grass or groundcovers (Figure 4.28). In contrast to this, where mound profiles are eased and where edges are less defined and varied, or where planting is more complex with shrubs and trees in non-defined patterns, the mound will appear much more naturalistic (Figure 4.29). In fact, in such cases, the mound may not be seen as a mound at all but give the impression that the transport corridor has proceeded into a cutting. The use of mounding on both sides of a transport corridor is often referred to as a false cutting.

The angle of a barrier not only has an effect on the direction of the noise it is reflecting, but it also has important aesthetic effects. If a barrier has a vertical profile, it appears to be a wall or a fence. If the barrier is angled

**4.26** A three-step ending of a transparent barrier is visually pleasing

**4.27** An architectural full stop applied to an architectural barrier

or curved, it suggests that it is something else, that it is a noise barrier (Figure 4.30). Moreover, viewing a vertical barrier from within a road corridor usually gives a sense of confinement. This feeling is eased by the angling of a barrier outwards, which tends to open up the view above and beyond. This impression also depends on the form, colour and top profile of the barrier. In order to reduce this sense of confinement it is recommended that the top profile of the barrier is kept simple and it is further improved if the top section is transparent. Viewing an angled barrier closely from outside a road corridor, where the top of the barrier is angled towards the viewer, may also present a feeling of confinement and claustrophobia. For this reason, it is recommended that, should there be public access immediately to the rear of a non-transparent and large barrier, it should not be angled. If, however, the

**4.28** Simple ornamental planting provides definition and character to a mound

**4.29** Naturalistic planting integrates an earth mound into the rural landscape

barrier is not immediately adjacent to public access and the barrier is small (human scale) and transparent, then angling of the barrier may be considered (Figures 4.31 and 4.32).

On the whole, angled and contoured barriers tend to have a more dynamic and designed appearance and tend to make more of a visual statement (Figures 4.33 and 4.34). In fact, most European barriers constructed over the last decade are angled. This creates an enhanced visual effect not only on the working side of the barrier but also at the community side of the barrier and even in broad landscape views (Figures 4.35 and 4.36). Some barriers have been angled out of physical necessity. Most angled barriers are tilted 3°–15° from the vertical depending on the materials and supports used, the height of barrier and the local surroundings. An example of this is the TGV rail line in Paris where a continuous barrier has been required: the ingenious use of intermittent angling allows gantries to be accommodated harmoniously into the design. This concertina effect adds important visual

**4.30** The complex profile of the barrier indicates that the barrier is more than a fence and that it is an architectural feature. Photograph courtesy of ONL [Oosterhuis_Lénárd], Rotterdam, the Netherlands

complexity to the barrier in an already complex visual setting and creates a strong design statement (Figure 4.37).

Contoured or curved barriers appear more pleasing when the curves are subtle. This is evident in the massive concrete barrier with its similarly profiled viaduct sections on the A16 motorway on the southern outskirts of Rotterdam (Figures 4.38 and 4.39).

Cantilevered barriers are designed in a range of shapes and sizes depending on their function (Figure 4.40). In most cases, however, a cantilevered barrier will reduce visual impact from outside the road corridor by easing or blurring the top edge of the barrier when viewed against the sky. Again, this can also be achieved by using transparent sections at the top of the barrier (Figure 4.41). Cantilevering also can effectively reduce the height of barriers by moving part of the barrier closer to the noise source. These overhangs may be extremely large and bold, but in many cases they are visually quite subtle. The decision to use a cantilevered barrier is usually based on acoustic as well as visual or aesthetic criteria, i.e. keeping the barrier as low as possible to reduce visual impact from outside the road/rail corridor while also reducing noise. A cantilevered barrier also appears more dynamic from within a transport corridor, especially where that corridor includes bends or curves. Cost is also an important consideration as supports and foundations may need to be more robust and complex than with other environmental noise barriers (Figures 4.42–4.45).

**Support structures, transitions and foundations**

As with any architectural form, the structure of a barrier may be used as a functional as well as an aesthetic element. The structure may then have an

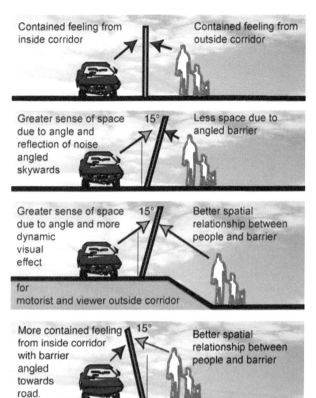

Contained feeling from inside corridor

Contained feeling from outside corridor

Greater sense of space due to angle and reflection of noise angled skywards

15°

Less space due to angled barrier

Greater sense of space due to angle and more dynamic visual effect

15°

Better spatial relationship between people and barrier

for motorist and viewer outside corridor

More contained feeling from inside corridor with barrier angled towards road.

15°

Better spatial relationship between people and barrier

**4.31** Angled barriers and spatial effects

**4.32** Angling of barriers towards people is acceptable when the barriers are small and transparent

**4.33** Angled barriers appear dynamic

**4.34** Angled barriers sit better in urban and semirural landscapes

important visual function, which helps to define the character of the barrier. This emphasis on the structure can vary from a minimal statement to a bold expression where the structure itself becomes the dominant visual element (Figures 4.46 and 4.47). Supports can also be appropriate or inappropriate in the context of their surroundings (Figure 4.48). In the case of gabion barriers, the structure itself forms the barrier (Figures 4.49 and 4.50).

Like many of the other elements of the barrier, the structure that supports the barrier requires the consideration of all of the design team. It is the structure supporting the barrier that can be subjected to significant stresses, especially wind loading, and must be designed to meet the

**4.35** An angled barrier is usually visually successful from the road/rail side where it is protecting land from development from rail and road traffic

**4.36** The community side of the barrier is also visually successful when the barrier is angled

appropriate engineering requirements. However, the design of the structure is also an important aesthetic consideration. It can either be emphasised, where the noise-reducing elements are barely visible and the structure itself is dominant (Figure 4.51). Alternatively, the structure can be downplayed or concealed within the façade to give a seamless appearance to the barrier (Figures 4.52 and 4.53). However, it should be borne in mind, when

**4.37** Splayed barrier makes a virtue out of an engineering necessity

**4.38** Viaduct section of an environmental noise barrier illustrating convex supports which tie into the overall convexity of the barrier

considering long barrier sections, seamless structures may appear monotonous unless other features are used to alleviate the uniformity of the barrier surface. A continuous undifferentiated surface may also simply be produced by using reinforced concrete produced within a rolling formwork. Potentially this can make for a visually blunt and heavy barrier which offers a tantalising surface for graffiti artists. Where there is a risk of graffiti it is prudent to use profiled or textured surfaces and to use the structure to break up any regular surface (Figures 4.54 and 4.55). A concrete barrier on the outskirts of Rotterdam, illustrates this point. The middle section of the modular concrete barrier, which is the most vulnerable to graffiti in terms of reach and visibility, is deeply profiled. The location of these absorptive panels also maximises the noise-reducing value of the barrier (Figure 4.56). The community side of the barrier is, however, more inviting, being smooth, but adjacent dense vegetation has deterred graffitists (Figure 4.57).

**4.39** Main part of the environmental barrier illustrating the overall convex profile of the barrier including the escape door section

**4.40** Large cantilevered barrier utilising a contemporary arm structure at Chiasso, Switzerland. Photograph courtesy of Greg Watts

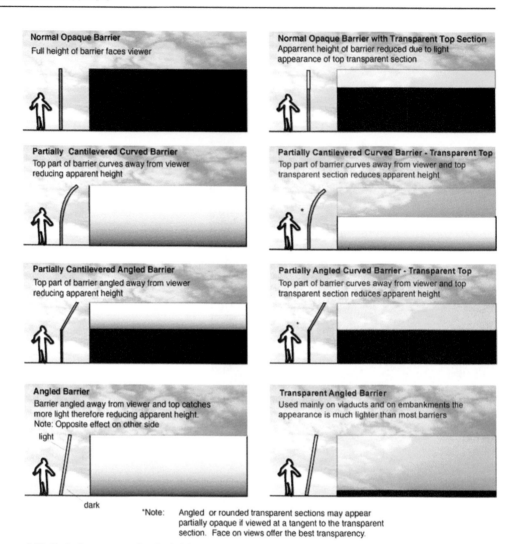

**Normal Opaque Barrier**
Full height of barrier faces viewer

**Normal Opaque Barrier with Transparent Top Section**
Apparent height of barrier reduced due to light appearance of top transparent section

**Partially Cantilevered Curved Barrier**
Top part of barrier curves away from viewer reducing apparent height

**Partially Cantilevered Curved Barrier - Transparent Top**
Top part of barrier curves away from viewer and top transparent section reduces apparent height

**Partially Cantilevered Angled Barrier**
Top part of barrier angled away from viewer reducing apparent height

**Partially Angled Curved Barrier - Transparent Top**
Top part of barrier curves away from viewer and top transparent section reduces apparent height

**Angled Barrier**
Barrier angled away from viewer and top catches more light therefore reducing apparent height.
Note: Opposite effect on other side
light

**Transparent Angled Barrier**
Used mainly on viaducts and on embankments the appearance is much lighter than most barriers

dark

\*Note:   Angled or rounded transparent sections may appear partially opaque if viewed at a tangent to the transparent section. Face on views offer the best transparency.

**4.41** Reducing apparent barrier heights

Despite the increase in the range of barrier types, for example the use of planar glazing, most smaller-scale barriers are formed from panels which are positioned directly between vertical posts. These posts are either fixed below ground onto piles or set within or bolted onto a concrete foundation (Figure 4.58). The number of vertical supports can be minimised by using lateral support structures either above or behind the barrier (Figure 4.59).

The use of posts can be avoided altogether in smaller barriers in which precast panels are bolted directly onto the foundations (Figure 4.60).

There has been an increase in the use of planar glazing and 'curtain walling' to mitigate noise (Figure 4.61). (Refer also to the section on Transparent barriers in Chapter 5). The advances in planar glazing systems, which originate in the design of building façades, have been utilized to good effect on four

**4.42** Elegant cantilevered barrier 20 m high and 18 m wide protects residential tower blocks in Hong Kong

**4.43** Public open space preserved by using a cantilevered barrier, Hong Kong

**4.44** Dynamic appearance achieved by cantilevered aluminium barriers, Bellinzona, Switzerland

**4.45** Substantial pre-stressed concrete support structure for cantilevered barrier, Dordrecht, the Netherlands (see Figures 4.59, 5.77 and 5.78)

**4.46** The structure is the dominant visual element, Gouda Railway Station, the Netherlands

housing estates in Vienna. The barriers use large glass panels that are supported by a series of posts, arms and tensioned cables (Figures 4.61–4.64).

Where more than a single or uniform material is used within the vertical or horizontal profile of a barrier, the transition is also an important visual and aesthetic consideration. Even when very different materials are used, a harmonious transition may be achieved by carrying elements from one section of the barrier into the other. On the traffic corridor side these transitions tend to blur because the barrier is viewed at speed. Yet this transition can be used as a design element and, if it is not considered as such, the visual quality of the barrier may be degraded (Figure 4.65).

The ground itself, and its geotechnical make-up are important considerations. Barrier foundations may need to be substantial, if, for example, a barrier is positioned on a road embankment, which may be of made-up

**4.48** The scale and design of concrete posts appear inappropriate against a natural backdrop

**4.47** The bold structure is incorporated into an overall design concept

**4.49** A gabion barrier itself forms part of the structure

**4.50** A freestanding concrete 'wave' wall. The screen provides its own support

**4.51** The barrier structure is an important part of the visual character of the barrier

ground. In such cases, deep piles which reach below the made-up ground may be required, together with ground anchors, to support the structure (Figures 4.66 and 4.67). In some cases the bottom of the barrier is integrated with the upstand of the barrier itself (Figure 4.68). The foundations can thus add considerably to the cost of the proposals. The foundations are critically important as the barrier may be subject to extreme wind loading. They will also have to be integrated with the road drainage system.

Many barriers use a steel I-beam as the main support. In the past I-beams have usually been terminated at the top by simply cutting the beam

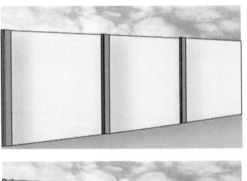

Barrier supports exposed on both sides

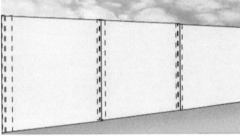

Barrier supports concealed on front side

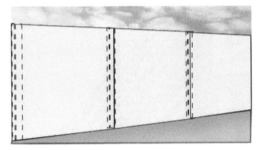

Barrier supports totally concealed

**4.52** Barrier support

**4.53** A seamless appearance of aluminium presents a high-tech image, A10, Amsterdam

horizontally. Sometimes these are capped (Figure 5.106). More contemporary-looking barriers, however, tend to cut the top of the I-beam at a slant. Visually this is more successful (Figures 4.69 and 4.70 view f).

**4.54** Graffiti artists find bland even surfaces irresistible

**4.55** Even well-executed graffiti will not enhance a barrier

**4.56** Profiled surface deters graffiti

## Fixings

Fixings are important elements, especially in terms of cost, weathering, maintenance and replacement. Fixings should allow the barrier to be fitted into place easily and allow for the easy removal and replacement of damaged panels or units. The issue of replacing panels is an important aesthetic consideration. It has been found to be worthwhile to manufacture and store additional panels where there is a risk that some of these may, over time, be damaged. Thus, when a barrier becomes damaged, it will be less costly to replace like with like. It will also be easier to maintain the intended appearance of the barrier. In the Netherlands, reductions in cost and the ease in which barrier parts may be replaced has focused the development of 'modular noise barrier' design and the development of a 'toolkit', 'containing a

**4.57** Dense undergrowth and paucity of passing people makes an inviting concrete barrier back less attractive as a graffiti target

limited number of different modular components'[2] (Figures 2.11 and 4.69). In most cases, fixings of regular noise-reflective or -absorptive panels comprise simple components (Figure 4.70). In some instances, the fixings require greater complexity, especially relative to the top section of the barrier (Figure 4.71).

## Other considerations

### Viewing at speed

The factor of speed affects the total design of the barrier. For the motorist or train passenger, the barrier is usually seen at speed, although the motorist has longer forward views than the train passenger, who has mainly close side-on views. In general, this means that the appearance of barriers and hence their design and construction needs to be simple, with clean lines and edges. Small nuances and changes in pattern, colour or texture are likely to be lost and in fact may create visual disharmony as they blur into a visual clutter. However, barriers are not only seen at speed; they often dissect areas in constant use by

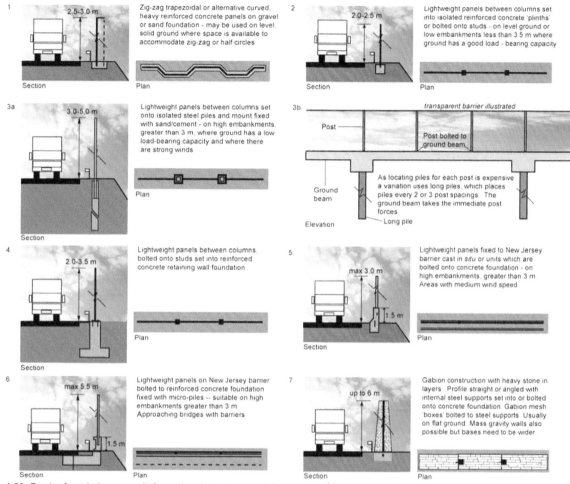

**4.58** Barrier foundation types (information derived, in part, from OECD[1])

local residents. The barrier design, therefore, needs to take account of their views as well as harmonising with the overall design and character of the urban fabric. In this respect, it is extremely important to stress that the barrier is not an independent structure which is designed only in its on right. It must tie in with the character of the transport corridor as well as the local surroundings. This is the most critical issue apart from the acoustic considerations. Each barrier, and each façade of the barrier, needs to be considered with regard to the local environment. This is evident in the design of the transparent barrier alongside the A27/A15 on-slip road at Gorinchem, south Holland. Here the identity of the transparent acrylic sheets placed between simple I-beam supports is transferred across the road bridge to create more of an identity and focus. The I-beams are cranked to create a diagonal line across the 'window', which contains an angled trough that acts as a gutter removing water from the barrier (Figure 4.72). The figures illustrate a number of important points of barrier design. Firstly, minor features such as the

**4.59** Steel lattice structure used to maximise the distance between support posts

**4.60** A modular panel from a postless absorptive granular concrete barrier system, Milan, Italy

angling and bending of the top of the acrylic sheet improves the visual effect. Additionally, the barrier components are repetitive and readily replaceable. Finally, Figure 4.73 shows the placement of concrete panels at intervals between the acrylic sheet panels. These have been located at regular intervals in order to limit the potential damage to the barrier length should part of the acrylic sheet be accidentally or deliberately set on fire.[3]

### Two-faced barriers

As noted earlier on in the chapter, it is not common for both sides of a barrier to look identical because it is usually only one side of the barrier that is required to reflect or absorb noise. The difference between the two sides of a barrier should be determined on aesthetic as well as acoustic grounds: each side of the barrier should be designed to integrate with the landscape character and the backdrop against which it is to be viewed. Thus, the face of the barrier that may be facing a road corridor may include a pattern or have a bright colour, whereas the other face, which could stand opposite housing, may be treated in a more discreet fashion. Here, the façade may be plain and designed to merge in with adjacent planting. Most barrier types can be designed with this in mind with the obvious exceptions of transparent barriers and many bio-barriers. Planting on either side of the barrier should also be designed with each separate identity in mind.

There are, however, many exceptions where road and rail corridors run side by side, which require both sides of the barrier to reduce noise. This may also occur where central reserve barriers are required between road carriageways. In these instances, the barrier, which is usually designed to absorb noise, may appear similar from either side. But, even here, both sides of the barrier need to be assessed relative to the surrounding landscape.

Another important consideration is the provision and maintenance of views into the landscape for motorists and rail passengers, and in this regard it may be appropriate to design barriers that allow views towards the landscape. This may conflict with the needs of those whom the barrier is intended

3  Galvanised steel 'I' beams on capped concrete plinth

6  Glass facade held in place with stainless steel plates that pass through and screw into the connector on the other side. The gap noted in note 5 occurs also with the vertical juxtaposition of the glass sheets - Evident in Photos 4 and 5

2  Community side facade of barrier - all supports and fixings and an additional tension bar located this side - construction profile is very shallow (except tension bar)

5  'I' beam connecting arms and stainless steal fixings on community side - Note the rubber pads which protect the glass on either side. Note the small gap between glass sheets is often filled with a rubber seal but this would create a visually obvious line between each sheet of glass

1  Street side facade of barrier with flush glass panels - all fixings and supports to rear except for tension bar at top

4  'I' beams with bolted side arms and fixings and capped concrete plinths

**4.61** Planar glazing at a housing project in Vienna

**4.62** Planar glazing protects housing and courtyards from the noise of a busy Viennese street

**4.63** Rooftop-level view of a 20 metre tall glazed curtain wall hung between the end façades of a U-shaped building

to protect (Figure 4.74). The new 'Betuweroute' that runs for a considerable length alongside the A15 motorway, providing a new freight rail link from 'Europort' Rotterdam to the German border, is an exception, as the rail line only carries freight. The noise mitigation is thus only concerned with the containment of noise within the corridor itself. The environmental noise bar-

**4.64** Curtain walling protects residents from road and rail noise in inner-city Vienna

**4.65** The repetition of vertical and horizontal design elements is used to achieve a harmonious transition between two systems

riers thus comprise extensive areas of cantilevered concrete noise absorptive panels, with some transparent sections above that contain views to and noise from the rail traffic. Although it was not necessary to consider the views from within the rail corridor, there appears to have been little consideration given to views from the protected community (Figure 4.75).

Although the research for this publication has not revealed a barrier incorporating one-way glass, this may be an appropriate solution in situations where views out from the transport corridor are desirable and where views into the transport corridor may be unwanted. The use of one-way glass may also be desirable in inhibiting bird strikes as the glass, from one side at least, appears solid to birds. (See Chapter 5 for more information on transparent barriers and bird strikes.) The containment and revealing of views can also be achieved by assessing the major views out and towards the corridor and then by designing

**4.66** Construction phase of a large transparent barrier showing extensive concrete piles

**4.67** Exposed concrete footing and bolt connections on a test barrier

kick-back windows or staggered panels which allow glimpses out but which restrict views in (Figure 4.76). Figure 4.77 shows how staggering can greatly reduce the visual impact of a barrier and allow views out of the road corridor, but unfortunately in this case the views are directly into a private garden.

**4.68** Transparent sections of a barrier fixed on top of the upstand part of a foundation

### Barriers: vertical or horizontal landscape elements?

Generally, barriers may be considered as horizontal elements in the landscape since the horizontal dimension is usually much greater than the vertical. In some cases, however, the vertical dimension is just as impressive as is evidenced by the 20-metre-high barriers which protect housing areas alongside the Périphérique in Paris and also various routes in Vienna (Figures 1.1, 1.2, 4.61–4.64).

Emphasising the vertical elements of a barrier may be an appropriate way of making a visual statement and a means of creating visual continuity and rhythm. However, if used in the wrong context, they can appear out of place or character and be visually intrusive especially if viewed against a natural backdrop. Cute and visually obvious solutions can appear trite and dated. Vertical elements can be used to fulfil a real visual function, for example, by directing attention to the location of the escape doors, but, even in such cases, this can be overdone (Figure 4.78).

### Repetition

Barrier design is generally based on the repetition of panels and structure (Figure 4.79). This strategy is important in keeping costs down and providing visual continuity, but with some barriers it can risk creating or introducing visually boring elements into the landscape. Such barriers can also alter the landscape character of an area and diminish landscape quality. Most barriers designed today follow one or more of four principal design approaches:

- to provide a simple barrier with no variety or material differentiation, or focal points along its length (Figure 4.80);

Modular noise barrier - woodland areas

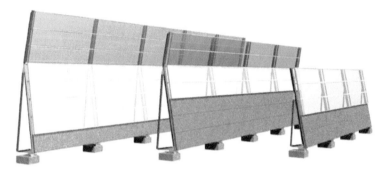

Modular noise barrier - meadow or other areas

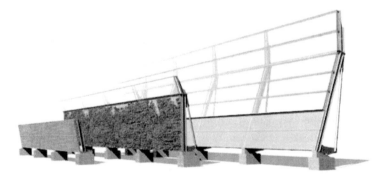

Modular noise barrier - variety of heights and panel configurations

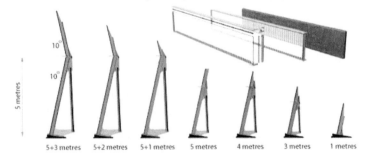

Modular noise barrier - support and panel configurations

**4.69** Modular noise barrier components in the Netherlands

**4.70** Simple fixings attach concrete panels to the supporting posts: (a) Road façade of a barrier with concrete and transparent panels; (b) View of community side of same barrier; (c) Concrete panels fixed to I-beam columns with steel plate flanges; (d) The void is filled with flexible rubber or foam spacers; (e) Road façade of a cementatious board and transparent barrier; (f) View of the community side of a barrier; (g) Simple fixing of bottom and middle panels to steel posts; (h) Middle panels that are located within an aluminium frame fixed with bolts and flanges

**4.71** More complex fixing systems: (a) Hierarchy of transparent panels along main route and slip road; (b) Section showing I-beam steel support and 'window' frame; (c) Bottom aluminium fixing trays to support transparent panels; (d) Top fixing of transparent panel; (e) Concrete panel barrier with transparent top section 'kick-back'; (f) Top section showing two part arrangement; (g) Detail view showing fixing and stabilisation of transparent elements

**4.72** Barrier design altered at a key focal bridging point along the length of the barrier

**4.73** Main extent of the barrier intersperses concrete panels between acrylic sheets. These have been added as 'firebreaks'

- to provide periodic focal points, features or periodic variations along the length of the barrier (Figure 4.81);
- to allow for a pattern within the overall façade of the barrier, such as staggered windows or a change in material (Figure 4.82);
- to provide visual variety by moving the horizontal alignment of the

View from the road/rail line:
Transparent barrier allows views to the landscape creating interest for drivers and passengers - in this photograph: sky, skyline, trees.

View from outside the road/rail line:
Transparent barrier allows views to the road/rail traffic which may be disturbing to local inhabitants and amenity users.

View from the road/rail line:
Opaque barrier stops views to the landscape creating and enclosed transport corridor insulated from the interest of the landscape.

View from outside the road/rail line:
Opaque barrier stops views to the road/rail traffic creating a more peaceful character. But in some areas the movement of traffic may add visual interest. There may also be long views of interest which are blocked although the near distance views of traffic are mitigated.

**4.74** Some advantages and disadvantages of transparent barriers

barrier away from and towards the transport corridor, by creating dog-legs, curves and splays.

A subtle, but sophisticated, example of this type is the competition-winning barrier along Route 16 in Copenhagen and its later variations (Figure 1.11). The arrangement of the barrier fixings to the upright supports enables changes in horizontal alignment as well as the vertical alignment of panels (Figures 4.83–4.85). Most barriers do not have this inbuilt flexibility and moving the barrier off its optimum line is rarely used, as the best location for the barrier is as close as possible to the noise source. It should be noted that a balance should be sought between the optimum location for noise and the most favourable location with regard to integrating the barrier into the landscape.

**4.75** The 'Betuweroute' freight train corridor from Rotterdam to Germany, alongside the A15 motorway. It is screened by curved concrete panels and in some areas topped with transparent sections

1. Straight barrier:
Transparent sections allow views, opaque sections prohibit views

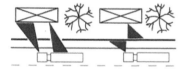

2. Staggered barrier:
Transparent sections allow views, opaque sections prohibit views

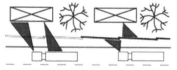

3. Zig-zag barrier:
Transparent sections allow partial views, opaque sections prohibit views

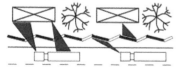

4. Straight barrier with occassional windows:
Transparent sections allow partial views, opaque sections prohibit views

5. Staggered offset barrier:
Transparent sections allow views, opaque sections prohibit views

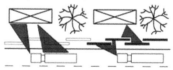

6. Staggered stepped barrier:
Small transparent sections allow partial views, opaque sections prohibit views

NOTE:
Barriers with transparent sections can benefit drivers and passengers by allowing views out but may disadvantage local people.

Barriers 3,4 and 6 limit views to the traffic corridor but allow some views from the corridor thereby creating interest for the driver and passengers.

**4.76** Barrier arrangements – opacity versus transparency

**4.77** Staggered pvc barrier with intermediate windows at right angles opens up views for the traveller

### Pattern

Applied patterns or patterns of light and shade created by barrier elements are an important tool in the designer's palette, but pattern should be used with due care. Patterns, if too simple, can appear stark, contrived and even puerile. Patterning, where required, needs to be sophisticated and bold (Figures 2.13, 4.63 and 4.86). If it is too subtle it may be lost when viewed at speed. Pattern should also be an integral part of, and relate to, the form and elements of the construction; surface treatments and painted patterns rarely appear successful. Patterns which are created in bas-relief work with light and shade and are usually more visually sophisticated (Figure 4.87).

Patterns are applied and created in a number of ways. On transparent barriers, bird-deterrent white lines and other embellishments are created by applying decals, sand blasting or acid etching, silk-screening or by locating the lines within the acrylic sheet (Figures 1.2, 1.10, 4.5, 4.12, 4.13, 4.23, 4.35, 4.61, 4.63, 4.84 and 4.86). Patterns on opaque panels can be cast (concrete) or embossed or applied (sheet metal). Explosive forming of sheet metal can be used to evaluate prototype designs[4] (Figure 4.88).

### Texture

The texture of a barrier is essentially defined by the material make-up of the barrier, although in many cases, when travelling at speed or when viewed from a distance, the textural nature of the material can appear indistinct. The texture of the material, therefore, should not affect the visual quality of the barrier except where close views are possible. Thus, for example, the porous nature of a concreted wood fibre barrier or an absorptive brick barrier is lost, except when viewed close up (Figures 4.89 and 4.90). Absorptive metal barriers generally appear as reflective metal barriers when viewed

**4.78** Pagoda structure draws attention to an escape door

**4.79** Barriers are usually made up of repetitive components as illustrated in this three-barrier hierarchy at Hardinxveld, the Netherlands

**4.80** A visually simple barrier on top of a viaduct section

from any distance. Timber barriers in the UK have increased in scale and many of the new barriers are noise absorptive. Despite this, they generally have a distinct 'garden fence' quality. In continental Europe timber is used in a variety of ways to provide a range of textures, colours and a play of light and shade (Figure 4.91). The textural effects of planting can also help to provide visual interest and complexity (Figure 4.92).

**4.81** A unique tongue-in-cheek feature incorporated within a barrier in Phoenix, Arizona. The mesh fence filled with river stone creates a visual symbol for a river while also visually breaking up the expanse of the wall

**4.82** An unusual arrangement of absorptive timber and brick treatments in a play of triangles at Altona, Hamburg, Germany

Furthermore, it must be emphasised that combining materials, textures and/or colours is of critical importance: usually a maximum of two should be used, not including that of the support structure. This may be increased to three, if planting is included as the third texture.

**Colour**

Colour is obviously one of the most important character-defining criteria for any architectural element and it is critical in determining the visual character

**4.83** A transparent barrier where the horizontal and vertical alignment of the panels can be altered by using flexible aluminium posts and arms. Photograph courtesy of Milewide A/S, Haderslev, Denmark

**4.84** Close-up view of barrier panels where the glass panels slightly overlap. Photograph courtesy of Milewide A/S, Haderslev, Denmark

and quality of the barrier. An interesting and challenging problem with colour in temperate zones is the changes of season. This is particularly an issue in rural locations where a colour chosen to blend into its surroundings in summer may

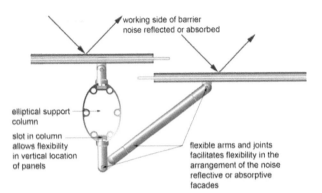

**4.85** Plan view of barrier showing flexible support arms allowing raising and angling of panels

**4.86** Extensive screening of apartments and open space areas. The addition of fish and bird decoration on the windows is likely to minimise bird strikes but will become visually dated over time

be overly conspicuous during the winter months. It is, therefore, not surprising that in many cases, the choice of colour is inappropriate. Colour, like any other design criterion, should be chosen for a particular reason and not be chosen arbitrarily, or simply because it is available. Colour, as with texture can also be determined by the material itself, although this can be changed through the application of paints, stains and anodising. A metallic colour, such as the silver sheen of aluminium gives a high-tech engineered appearance. Muted green colours and timber colours give the effect of nature and the natural environment. Bright colours, such as reds, yellows and oranges imply conscious design and the making of a statement. Thus, colouring parts of a barrier red, for example, states that the barrier has some visual/architectural merit and in such cases care should be taken to make sure that this statement is justified. Bright colours are used to both good and bad effect in urban and suburban locations, but they are rarely successful in rural locations, It may be a function of present-

**4.87** Visual interest created by a bas-relief pattern

**4.88** Butterfly motif created by 'explosive forming' in aluminium

day fashion, but it appears that in the UK and the rest of Europe, muted non-contrived colours and tones appear to work well in most urban locations. Whites and off-whites, concrete shades, greys, the light and shade of glass, metallic sheens and muted colours appear to fit in and are easy on the eye. An exception, where bright colours do work visually, is found where the total design is well conceived (Figure 4.93). In rural and semirural locations muted tones also appear pleasing. Glass, timber tones, tawny browns, olive, oaky greens and some greys appear tranquil. It is also surprising to note that some darker russet and burgundy tones appear satisfactory and do not jar the eye as might be expected (Figures 4.94 as well as 4.39, 4.56 and 4.70).

Care should be used when selecting coloured transparent elements. Slight tints of grey and blue usually appear unsatisfactory as they distort the colour of the sky and may not fit in well with the surrounding colour of the vegetation (Figures 4.95 and 4.96). Rose, yellow and beige tinted glass can appear dull, lifeless and out of place in the landscape. However, the occasional use of brightly coloured glass can be a very effective design element, adding

**4.89** Textural detail of an absorptive brick barrier

**4.90** Texture becomes a less significant factor with distant views

**4.91** Colour, form, light and shade combine to add interest to a timber absorptive barrier

**4.92** Planting helps to provide visual interest

interest and breaking visual monotony (Figure 4.97). Additionally, the occasional use of a bright colour, which highlights a specific architectural feature, can lift the appearance of the overall design (Figure 4.98).

In many instances, colour and tone are graded from dark at the bottom of the barrier to light at the top, where it may be viewed against a lighter sky. This is an effective design solution, especially where the colour gradations are balanced and subtle. Although some bold barriers use contrasting colour changes to create a confident visual statement (Figures 4.39 and 4.56), ambitious colour changes may appear to be stark, ill-fitting and contrived unless they are part of a visually robust and complex urban setting (Figure 4.99). Reflections, particularly with transparent barriers, should also be considered as they can change the character of the barrier (Figure 4.100).

In Switzerland, Swiss Federal Railways has chosen a particular dark grey colour for all of its absorptive barriers. This colour is effective in that it reflects, to a great extent, the colours of the railway environment, including the greys of the steel tracks, rolling stock, gantries, concrete sleepers and the granite ballast that holds the sleepers and tracks in place. The colour was chosen for all barriers so as to provide a particular corporate image and thus, when such a barrier is seen, it says 'Swiss Federal Railways'. In many respects, the chosen colour is correct and the corporate identity argument appears logical. However, the factors of light and shade and the character of individual locations have largely been ignored. This undoubtedly means that, although some barriers fit neatly and sensitively into the environment, others do not (Figure 4.101).

The creation of route identity is also a driving factor in the design and colouring of barriers for the 'Betuweroute' rail freight route between Rotterdam and Germany. The railway noise barriers are light concrete coloured on the outside and inside except where absorptive panels are used. These are dark grey (Figure 4.102).

Colour can be used to great effect to focus attention in the landscape. Two projects, both using red, combine dramatic colour and massive form to

**4.93** Well-conceived use of strong colours

**4.94** Burgundy tones work surprisingly well

create remarkable identities adjacent to the regular motorway landscape. The Brembo Research Office located alongside the Milan–Venice expressway at Bergamo in Italy uses an 'outsized' wall to 'punch' the noise away from the buildings and open space areas with a 1,000 metre long 'red line', while reinforcing the dynamic identity of the brake manufacturer for Ferrari and other notable marques. The barrier utilises gleaming grooved, lacquered aluminium panels, next to a red car park. (Figure 4.103). The first phase of the 'Wall' alongside the A2 motorway at Utrecht uses a dynamic, red, twisting and flowing form located above transparent sections. The flowing profile will eventually form part of the fascia and roof structure to a shopping and restaurant complex that will protect the proposed new mixed development located behind it (Figure 4.104).

### Light and shade

The fall of light onto the barrier and shade created by angles, appendages, material make-up, posts and supports, and vegetation constitute an important visual element and are as important as colour or texture.

In choosing a barrier and its overall design and colour, it would be well to note the aspect of the barrier and whether it will receive sunlight or whether it would be in shade or partial shade. Thus, a barrier which runs alongside a north–south transport corridor may well receive bright sunlight on one side and be in deep shadow at different times of the day, whereas a barrier located

4.95 Older-style transparent barrier using grey-blue transparent panels. Red posts enhance the old bridge structure

**4.96** Well-intentioned blue transparent barrier in a dominantly green landscape is slightly jarring

**4.97** Spots of colour lift the overall design

alongside a corridor on an east–west alignment may have one face always in shadow. The colour of the side of the barrier in perpetual shadow may need to be lighter to achieve the desired visual effect.

Light falling on a barrier can also create pleasing effects, which alter according to the strength of the changing light, the weather, the angle of the sun, time of day, etc. This allows a barrier to become more visually complex and interesting, especially when placed in urban locations where views are most likely

**4.98** Striking colour used as part of an overall design concept

**4.99** Robust colour used to enhance a visually complex urban setting

to be from close up and for longer periods of time. Visual complexity is important in these locations in order to avoid visual sterility (Figure 4.105).

### Profiling

Profiling, or the repeated raising and pushing back of the barrier surface, is an important consideration for absorptive barriers, as the increased surface area of the barrier can enhance sound absorption (Figures 4.106 and 4.107). Profiled barriers also have the advantage of making it difficult for graffiti artists to paint coherently and therefore they are dissuaded from using surfaces which are irregular. Profiled barriers are also visually more interesting because their appearance changes throughout the day. Light highlights the peaks of the profiling at some times of the day, while causing shadows in the troughs (Figure 4.108).

Profiling of a barrier in terms of angles, concavity and convexity affects

**4.100** The 'Cockpit': Reflections of clouds and other surrounding features form part of the visual character of the barrier at different times of the day and relative to weather and daylight conditions. Photograph courtesy of ONL [Oosterhuis_Lénárd], Rotterdam, the Netherlands

**4.101** Swiss Federal Railways corporate barriers protecting an urban environment

noise reflection as well as visual character. Whereas a concave barrier will reflect noise back towards the source, a convex barrier may be used to reflect noise skywards. Barrier supports can be used to enhance the visual quality of the profile (Figure 4.38).

### Materials and design

Historically, noise barrier designers have often used the palette and techniques of the architect. A barrier was seen to be a façade, just like any façade of a building. Today, however, many barriers have a particular character that allows their form to integrate with their function. Contemporary design attitudes dictate that verity is applauded and artifice is derided, so barriers appear to be more honest if they do not try to hide their function through trite patterning or *trompe l'oeil*. Where a noise barrier is likely to be

**4.102** The 'Betuweroute' utilises light-coloured concrete panels inside and out except for areas where absorptive panels are required. These are dark grey

a visual feature in the landscape it should be designed with integrity and regard to its form and function, although over time this fashion may well change. Barriers should thus be designed using the best expertise, technology and materials available, so that they fit well into the 21st-century *oeuvre* and will stand the test of time, at least for their design life of 40 years.

The two main reasons why barriers appear incongruous or characterless in the landscape or in urban locations are function and cost. Although there is no doubt that function is the primary reason for building a barrier, this should not be done to the detriment of the landscape and visual context. The function of reducing noise and/or visual intrusion must be seen in the context of the overall visual effects and impacts. In the UK, all too often it appears that off-the-shelf products are used provided they fulfil the task of reducing the noise to the specified level. Some design may take place, but this does little to reduce the visual effects caused by the selection of inappropriate materials and does little to maintain and protect the quality of the environment. Even if token screen planting and mounding is provided, it cannot disguise a poor design. Barriers should only be designed once studies of the local natural environment have been undertaken, including studies of materials, built forms and massing, colour and vegetative patterns and the historical context. Once this is completed, concepts should be explored that explain the designers' reasons for the visual appearance of the barrier.

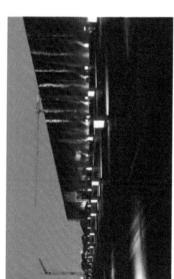

**4.103** A red integrated barrier illustrated in concept and as built provides a conspicuous and confident statement at the Brembo HQ outside of Bergamo, Italy. Images and photographs courtesy of Ateliers Jean Nouvel, Rome, Italy

**4.104** 'The Wall': A red integrated barrier illustrated in concept and during construction provides a new shopping centre which screens proposed development behind it. Images and photographs courtesy of VVKH Architecten, Leiden, the Netherlands

**4.105** The cast of light and shade adds visual complexity to galvanised steel columns

**4.106** Zigzag profiling of a barrier on a viaduct in Switzerland increases visual interest and noise absorption. Working side shown during construction. Photograph courtesy of Greg Watts

**4.107** The community non-working side of the barrier also creates visual interest for the public away from the barrier. Photograph courtesy of Greg Watts

The arrangement of planting, maintenance and safety zones alongside the barrier also needs to be considered, since this contributes to the overall visual impression. These zones may, of course, overlap and widths may vary according to the location and the type of barrier or road (Figure 4.109).

For acoustic reasons, barriers are generally positioned as close as possible to the noise source so as to be most effective. However, visually this is not the

**4.108** Profiling adds interest to a potentially bland surface

case as it is preferable in most locations to provide space on both sides of the barrier. These spaces perform a number of visual functions. Firstly, distance allows the driver/traveller/pedestrian to be separated from the barrier, thus avoiding or reducing a feeling of claustrophobia. Distance also reduces the apparent scale of objects and, most importantly, space gives the scope for planting, which is a crucial design element. The use of these spaces can conflict with acoustic design objectives and it is necessary to reach a compromise (Figures 4.110 and 4.111).

Setting a road within a cutting is a technique often used to reduce noise. However, additional screening may be required to achieve the noise objective. This would usually involve placing a barrier at the top of the cutting slope, but this could negate the visual benefits provided by the cutting. Furthermore, it can exacerbate the visual impact by silhouetting the barrier against the skyline. This again is an area where a compromise between the acoustic and landscape objectives will need to be reached.

Cost obviously plays a large part in dictating overall design concepts. However, costs should be properly balanced against environmental and visual impacts, and the potential effects on the quality of life. It must be remembered that barriers will be part of the landscape for a long period of time. It is important that these structures and the materials they comprise stand the test of time. They will not do this if the visual quality of the barrier is sacrificed for cost-cutting reasons.

### Choosing materials – visual neutrality and compatibility

One way of choosing materials for a barrier on aesthetic grounds is to link the character of the landscape/townscape in a 'neutral' way with the character of the materials. Thus, for example, in rural agricultural areas where the predominant character may comprise the earth itself, grass and native trees and shrubs, it makes sense to integrate the barrier with an earth mound and with grass, trees and shrubs, which are visually and materially neutral. Where, for example, a barrier is required to pass through a woodland, it can be appropriate for it to be designed using timber or other organic materials, which are materially and visually sympathetic with the character of the environment. Transparent barriers which are visually neutral also tend to be visually effective in rural landscapes.

Alternatively, if the landscape character is dominated by the route corridor

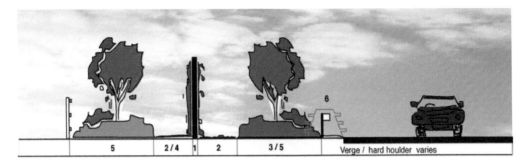

Horizontal alignment : Potential Arrangement 1

1.    Barrier with climbers.

2.    Maintenance zone - may vary depending on type and height of barrier - usually at least 1m on either side. May be soft or hard surface depending on location. May require drainage below ground. Climbers may be included in this area.

3.    Safety zone to the front of the barrier. Safety fence may be required depending on the vertical alignment of barrier and the distance of the barrier from the carriageway and the road standard.
Steps over safety fence may be required at periodic intervals and at escape doors.

4.    Safety zone to the rear of the barrier to allow travellers to move out of the road corridor via an escape door.
This will probably  tie in with the maintenance zone.

5.    Planting zone - may vary depending on need to integrate barrier, height of barrier and materials and overall landscape strategy / concept and available space. Position of boundary fence varies depending on who is responsible for the maintenance of the planting. If a fence is required then this should be chosen to tie in with the overall landscape scheme.

6.    Safety fence may be required depending on vertical alignment of barrier and the distance of the barrier from the carriageway.

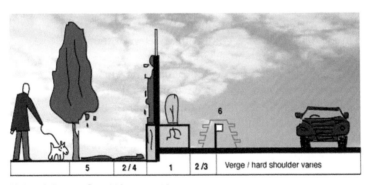

Horizontal alignment : Potential Arrangement 2

1.    Barrier with planter and hedge - planter with climbers and climbing frame or wires.

2.    Maintenance zone for barrier and escape zone from road behind safety fence.

3.    Safety zone to the front of the barrier. Safety fence may be required depending on vertical alignment of barrier and the distance of the barrier from the carriageway.
Steps over safety fence may be required at periodic intervals

4.    Safety zone to the rear of the barrier to allow travellers to move out of the motorway corridor via an escape door.
This will probably tie in with the maintenance zone.

5.    Planting to help integrate barrier into local surroundings.

6.    Safety fence may be required depending on vertical alignment of barrier and the distance of the barrier from the carriageway.

**4.109** Horizontal alignments and land use

itself, with road furniture and buildings clearly visible, the barrier is likely to be much more successful if explicit reference to these is made in the design and in the use of similar 'manufactured' nonorganic materials. An excellent example of where visual neutrality, form, function and aesthetics work hand in hand with the environment is located along the TGV line into Paris (Figure 4.112). Here, the materials chosen tie in with the overall character of the railway line which is dominated by steel in the gantries, cables, railway tracks etc. The grey

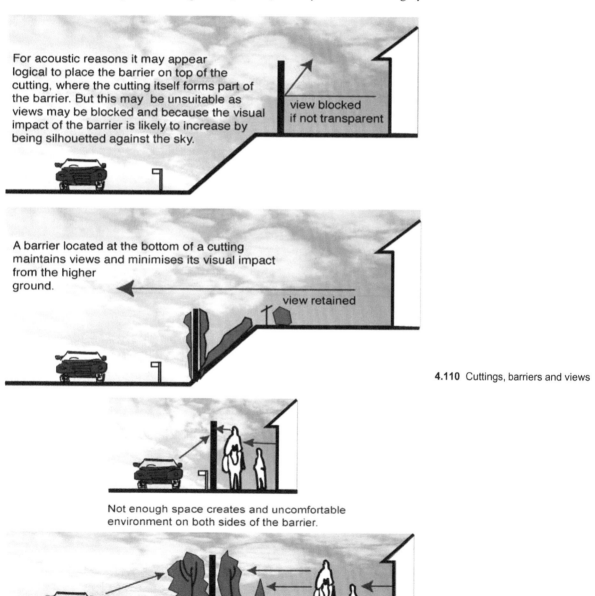

For acoustic reasons it may appear logical to place the barrier on top of the cutting, where the cutting itself forms part of the barrier. But this may be unsuitable as views may be blocked and because the visual impact of the barrier is likely to increase by being silhouetted against the sky.

view blocked if not transparent

A barrier located at the bottom of a cutting maintains views and minimises its visual impact from the higher ground.

view retained

**4.110** Cuttings, barriers and views

Not enough space creates and uncomfortable environment on both sides of the barrier.

Additional space removes barrier from personal space leaving the traveller and inhabitants feeling more comfortable and allows the appearance of the barrier to be softened over time through planting.

**4.111** Barriers and the need for space

pink colour of the steel barrier is balanced against the background colours of apartment blocks, which are generally pastel shades of pink, ochre and grey. (The combining of materials within a barrier is discussed in Chapter 5). Another project, this time in Australia, uses the colours and materials of the

**4.112** Well-chosen neutral colours and materials allow a large structure to integrate with its surroundings

**4.113** Corten steel, rust-coloured barrier winds itself aesthetically along and over the Craigieburn Bypass, Melbourne, Australia. Photograph courtesy of Peter Hyatt and Tonkin Zulaikha Greer Architects, Sydney, Australia

**4.114** Planting unifies the visual appearance across road and rail corridors

**4.115** Lighting brings a transparent barrier to life at night at the Craigieburn Bypass, Melbourne. Photograph courtesy of Peter Hyatt and Tonkin Zulaikha Greer Architects, Sydney, Australia

Australian soil as a direct reference. The Corten steel sheets of a barrier at Craigieburn, Melbourne are used in simple concave and convex folds to produce a gently undulating russet-coloured wave (Figures 4.113, 5.34–5.36).

The issue of compatibility becomes particularly important and more challenging in situations where more than one barrier is required. These situations occur where barriers are required on either side of a road corridor, when a central reserve barrier is used or where they are aligned between road and rail corridors. It is recommended that materials are co-ordinated carefully between the barriers and across the routes. This can be achieved by using barriers with similar colours, but often the principal method is to use planting as a unifying theme (Figures 4.114 and 4.79). Figure 4.79 illustrates the successful realisation of a limited palette of materials and colours.

### Night-time barriers

Lighting is usually an integral part of architectural and landscape design and yet few environmental noise barriers are designed to have any visual presence at night. This is for three main reasons. Firstly, for road safety, driver distraction needs to be kept to a minimum. Secondly, in most areas, residents would not want additional visual stimulus and, thirdly, there is a moratorium on the use of lighting that will increase night-time glow and light pollution. In most cases, if lighting is incorporated as part of the barrier design, it must have the following credentials:

* Lighting must not be too distractive to drivers so that it may cause accidents;
* Lighting must not increase light pollution and night-time glow;
* Lighting must be aesthetically pleasing for visual receptors both within the road corridor and those viewing from outside. These criteria have been met along one of the barriers along the Craigieburn Bypass, Melbourne, where a transparent barrier is subtly lit to create a visually pleasing translucent effect (Figure 4.115). The barrier panels are edge-lit acrylic, sandblasted with a digital pattern and overlaid with coloured precast concrete blades.

### References and endnotes

1.  OECD (1995) *Roadside Noise Abatement*, OECD, Paris.
2.  Innovatie programma Geluid voor wegverkeer, 'Scientific Strategy Document, DWW-2007-019', April 2007, p. 193.
3.  This follows on from the deliberate 'torching' of the cantilevered barrier along the A16 at Dordrecht, where the barrier was set alight by placing a motorbike adjacent to the barrier and then setting it alight.
4.  Explosive form originates in the defence industries. A metal sheet is located against a die and an explosive charge in a large water-filled tank. The explosion causes the sheet and die to be forced together creating a relief image. This process is relatively expensive and is mainly used to test patterns before dies are made for metal pressing on a larger scale. Refer to http://www.exploform.com/technology.php.

# Types of barrier and barrier materials

# 5

## Introduction

There are broadly three types of acoustic barrier, namely reflective, absorptive and reactive which would initially be selected for acoustic reasons, but this choice also determines the range of possible appearances. By their nature, absorptive and reactive barriers are always opaque, whereas reflective noise barriers may be opaque and act as visual barriers as well, or they may be transparent and lighter in appearance. Transparent barriers allow full or partial views through the barrier and light is not obstructed as is the case with an opaque structure. Frosting, colouring, silk-screening and other techniques are used to reduce transparency where views and light may need to be partially obstructed. These barriers also appear lighter in form, but may have the advantage, if required, of partially screening open views to traffic (Figure 5.1).

Sound-absorptive barriers contain a porous element that absorbs noise. This porous material can form the surface of the barrier, as is the case with concreted woodfibre and granular concrete barriers (Figures 5.2 and 5.55). Less robust absorptive materials such as mineral wool are protected and enclosed within a skin, where the side facing the noise is perforated. These casings are usually made from timber, steel or aluminium sheeting and brick (Figures 5.3 and 5.4 and 4.82).

Reactive barriers are those which incorporate cavities or resonators designed to attenuate particular frequencies of noise. Sound enters these cavities via small holes or slots in the face of the barrier. Block-work barriers incorporating such cavities have been in use for many years and have been to incorporated within a New Jersey safety barrier. This system has been patented and has been installed in the vicinity of Milan (Figure 5.5).

**5.1** Semitransparent noise barrier provides a visual screen while maintaining natural light along a footpath

**5.2** Deeply profiled cementatious porous façade partially absorbs noise and creates a play of light and shade

## *DMRB* guidelines

The Highways Agency has published guidance on the use of materials for barriers in its *Design Manual For Roads and Bridges*.[1,2] This guidance is intended for the design of noise mitigation for trunk roads and motorways but is also relevant to the design of barriers in other locations. Advice is given for a range of materials and, where appropriate, reference is made to the following Highways Agency design specifications and British Standards, which should be complied with:

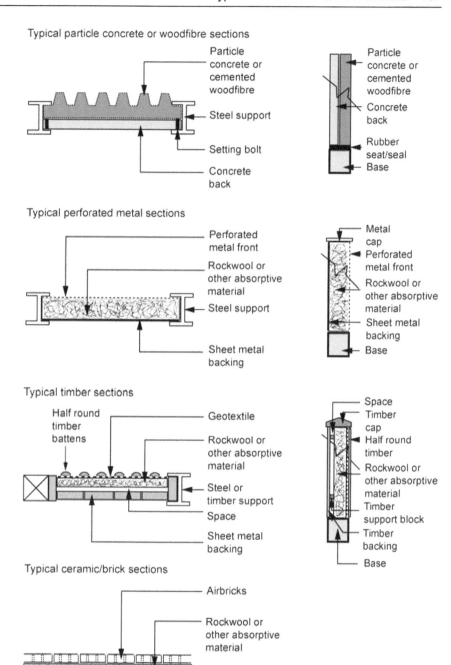

**5.3** Typical absorptive-type barrier sections

- Timber: *Specification for Highway Works* (MCHW 1) clauses 304, 310 and 311. Structural timber should comply with BS 4978, BS 5756 and/or BS EN 519 and/or BS 5268–2 and BS EN 1912 and in terms of sustainability comply with clause 126, BS 1722–7;

**5.4** Section of a sophisticated aluminium barrier with permeable front façade, internal space, mineral-wool-absorbing material and nonpermeable community-side façade

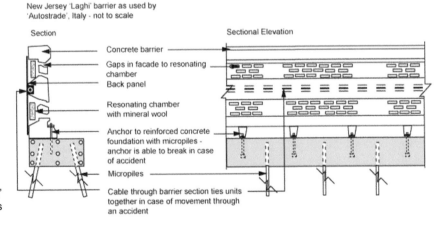

New Jersey 'Laghi' barrier as used by 'Autostrade', Italy - not to scale

Section

Sectional Elevation

Concrete barrier

Gaps in facade to resonating chamber

Back panel

Resonating chamber with mineral wool

Anchor to reinforced concrete foundation with micropiles - anchor is able to break in case of accident

Micropiles

Cable through barrier section ties units together in case of movement through an accident

**5.5** New Jersey 'Laghi' barrier with resonators

- Brick Walls: *Specification for Highway Works* (MCHW 1) Series 2400, BS EN 772 and BS 5628–1 : 2005, *Code of Practice for Use of Masonry, Code of Practice for the Use of Unreinforced Masonry*;
- Concrete: *Specification for Highway Works* (MCHW 1) Series 1700 and 2000, and BS 8110 Part 1, *Structural Use of Concrete* (to be replaced with Eurocode 2);

- Metal: *Specification for Highway Works* (MCHW 1) Series 1800, BS 5950 Parts 1 and 5, *Structural Use of Steelwork in Building*, and BS 8118, *Structural Use of Aluminium*;
- Transparent Materials: German standard ZTV Lsw-88;
- Reinforced Soil and Anchored Earth Structures: *Specification for Highway Works* (MCHW 1) Series 2500 'Special Structures';
- Environmental Barriers: *Specification for Highway Works* (MCHW 1) Series 2500 'Special Structures'.

For each material type, the advice lists many of the qualities and characteristics that can influence the appearance and design of a barrier. Durability is seen as an important factor and all barrier materials are required to remain serviceable for 40 years and require no maintenance for 20 years. Some potential drawbacks associated with each material are also listed, although no advice is given on ways of overcoming such problems. In addition to the range of manufactured materials described, it is also noted that vegetated barrier systems can be used to good effect, but no mention is made that these need regular maintenance.

The guidance is essentially a design checklist and it is clearly drawn both from experience in the UK and overseas. Although containing much useful information, it is not comprehensive. An example of this is found in its discussion of the treatment of transparent materials to deter birds from flying into a barrier. It is noted that materials may be tinted or a pattern of thin opaque stripes may be applied, but no mention is made of other techniques that are in use. In some countries, silhouettes of raptors are often applied to the barriers, but these do not enhance the appearance of the barrier in any way. They are also only effective if applied at a high density, which would be visually unattractive. The application of bird silhouettes should be avoided except where they form part of an overall design concept (Figures 4.5, 4.32, 4.35, 4.62, 4.64, 4.83, 4.84 and 5.6). Moreover, tinted glass should primarily be used as a design element and not to deter birds.

When dealing with the need for maintenance, it is noted that transparent barriers require more cleaning than other materials, but this is not necessarily the case if surface treatments are applied which inhibit the settling of dirt and render the barrier self-cleaning in rain. A number of self-cleaning products are

**5.6** Deterring birds using bird silhouettes appears naïve and diminishes the design integrity of the barrier. Unless densely applied they are also ineffective

available that have been incorporated into glass and paint. Other products may be applied to barriers that facilitate the easy removal of graffiti. Attention could also be drawn to ingenious solutions which have been adopted to facilitate cleaning. In Italy a collapsible transparent barrier has been installed near Milan which incorporates a hinge, allowing the inaccessible rear façade to be lowered and cleaned from the carriageway (Figure 5.7).

The *DMRB* advice on materials highlights many important issues associated with barrier materials and their maintenance. It does not attempt to offer guidance on the selection of materials for particular situations. Barrier designers in the UK will need to use the *DMRB* in conjunction with other advice when developing successful barrier concepts.

## Noise barrier types

### Earth mounds

Earth mounds, bunds, or berms as they are called in the USA, are often used to screen development and infrastructure projects. Indeed, earth mounding and the creation of false cuttings are ubiquitously found alongside motorways and trunk roads in rural, semirural and even in urban and suburban locations (Figures 4.28, 4.29 and 5.8–5.10). In the appropriate location, earth mounds have distinct advantages over other noise barriers in that they:

- may have a 'natural' appearance and may not appear to be noise barriers at all;
- may create an open feeling in contrast to a vertical or cantilevered screen;
- normally do not require additional safety fences;
- may cost less if excess material is available from construction;
- may be less costly to maintain;
- usually have an unlimited life span.

**5.7** Hinged transparent barrier allows the inaccessible rear to be cleaned

**5.8** A well-planned earth mound protects a suburban environment

**5.9** An earth mound used as an ornamental feature for a retirement home

**5.10** Earth mounds in rural areas can be used for habitat creation and ecological enhancement

Earth mounds may be effectively integrated into the local landscape context and, when they are planted or seeded with grass and/or wild flowers, they may form an attractive barrier which will in time be unrecognisable as a barrier in the landscape. Earth mounds do, however, require much more space than a vertical barrier. This is because the earth mound comprises a berm at the top and sloping sides and also generally needs to be higher than a vertical barrier to achieve the same acoustic performance (Figure 5.11). However, it should be noted that space may be needed for planting on each side of a tall vertical barrier to make it visually acceptable.

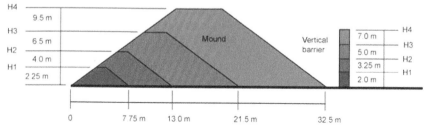

Comparison of heights in metres

|  | H1 | H2 | H3 | H4 |
|---|---|---|---|---|
| Vertical barrier | 2.0 | 3.25 | 5.0 | 7.0 |
| Mound | 2.25 | 4.0 | 6.0 | 9.5 |

NOTE: The dimensions will vary as berm widths and slope angles depend on the structural stability of different soils and fill materials

Not to scale

**5.11** Mound/vertical barrier height comparison (adapted from Roads and Traffic Authority of NSW[3])

As with any barrier, there are two sides to any mound or bund. The side slopes of either side of the mound may need to be treated differently. The side slopes of the mound are determined by a number of factors:

- the required acoustic performance of the mound;
- the geotechnical nature of the material make-up of the mound;
- the height of the mound relative to its topographical surroundings;
- the cost of the material that will make up the mound;
- the availability of space to accommodate the mound;
- the landscape character of the adjacent land and its land use and the appropriate angle of repose of the soil to help integrate it into the landscape context.

The severity of the slope may well be determined by the acoustic need to keep the face of the slope as close to the noise source as possible, which will in turn help to keep the mound lower than would be the case if the slope was eased. However, the more extreme the side slopes of a mound, the more difficult it is to construct and plant: in addition, planting would have more difficulty in becoming well established and may require more frequent and costly maintenance. Mounds with severe slopes also tend to look unnatural in the landscape, although in urban-scapes the matter of unnaturalness is less important. It is also worth noting that from the point of view of road or rail users the unnaturalness may not be a decisive factor as the route corridors themselves are hardly natural elements in the landscape. The rear side of these mounds and the interface with the natural landscape is, however, different and every effort should be made to integrate the mound as naturally as possible with its surroundings (Figure 5.12). It should be noted that many of the larger mounds in the Netherlands are now 'topped' with a small stone-filled gabion wall or timber fence. The topping of the mound with a hard

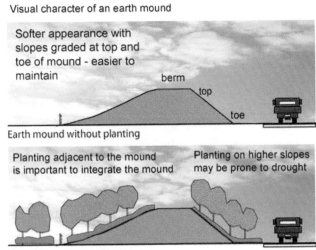

Visual character of an earth mound

Softer appearance with slopes graded at top and toe of mound - easier to maintain

berm

top

toe

Earth mound without planting

Planting adjacent to the mound is important to integrate the mound

Planting on higher slopes may be prone to drought

**5.12** Earth mounds – slopes and planting

Planting and an earth mound

**5.13** A large earth barrier with a gabion barrier on top improves noise mitigation due to the hard diffracting edge on top of the barrier

edge rather than a soft (earth) edge increases noise reduction as noted in Chapter 3, 'Bunds versus screens' (Figure 5.13).

The appropriation of land is a serious issue and there is pressure to keep mound gradients as steep as possible to minimise land-take. Expert geotechnical advice will be needed to determine the maximum slope that can be achieved to avoid soil slippage for the given soil characteristics and mound height required. In most cases, the maximum side slopes tend to be in the range of 1 : 2 to 1 : 3. All planting on slopes will need maintenance and the slope angle will determine the way that maintenance is carried out. With shrubs and trees, maintenance is usually carried out less frequently than with grass. Grass slopes can be maintained by mowing as shown in Figures 5.14 and 5.15.

In rural areas, it is usually more appropriate to minimise gradients on pasture to 1 : 6 or less to allow grazing. In arable areas, side slopes of 1 : 10

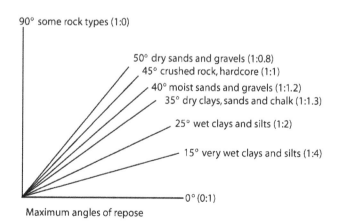

90° some rock types (1:0)

50° dry sands and gravels (1:0.8)
45° crushed rock, hardcore (1:1)
40° moist sands and gravels (1:1.2)
35° dry clays, sands and chalk (1:1.3)

25° wet clays and silts (1:2)

15° very wet clays and silts (1:4)

0° (0:1)

Maximum angles of repose

**5.14** Soil slope diagram (after the *New Metric Handbook*[4])

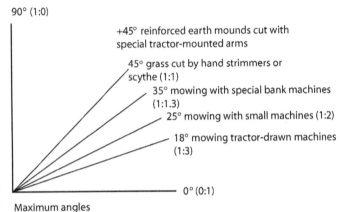

**90° (1:0)**

+45° reinforced earth mounds cut with
special tractor-mounted arms

45° grass cut by hand strimmers or
scythe (1:1)

35° mowing with special bank machines
(1:1.3)

25° mowing with small machines (1:2)

18° mowing tractor-drawn machines
(1:3)

0° (0:1)

Maximum angles

**5.15** Soil slope and grass cutting diagram (after the *New Metric Handbook*[4])

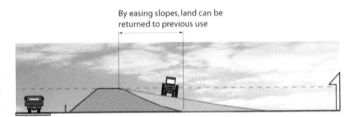

By easing slopes, land can be
returned to previous use

**5.16** Easing the slopes of a mound retains the same height but allows for a return to previous use

or less are appropriate to allow for mechanised farming. Where possible, the gradient of the mound should match the natural slope of the surrounding land. In this way, a greater proportion of the land may be returned to its original use (Figure 5.16). During construction, every attempt should be made to retain the existing soil characteristics, by appropriate soil stripping and storage, and by restoring it with appropriate laying techniques. Care must be taken when importing subsoil and topsoil that their characteristics are compatible with the existing soils. Notwithstanding, land which is disturbed during construction and returned to agriculture will not necessarily be of the same quality as its surroundings. The resultant disturbance of topsoil and soil structure may adversely affect drainage. For these reasons, not all farmers favour this solution.

Steep profiles may look engineered, especially until planting becomes established, which is more difficult on steep slopes than on shallow ones. To reduce the engineered effect, the profile of the mound may be varied along its length as this will help to give it a more natural appearance, which may be particularly important where the surrounding topography is undulating. The merging of the top of the mound into the berm and its toe into the ground also affects its character. In certain situations, a clean appearance may be what is required, but sharp transitions between horizontal planes and slopes are more difficult to maintain (Figure 5.11).

When deciding the grading of the mound, it is essential to keep the drainage characteristics of the mound and of the surrounding land in mind. Consideration must also be given to the fact that the upper parts of the mound will tend to be drier as water will move downwards towards the toe of the mound. Planting must then be designed according to this constraint.

Structures with steeper than 1 : 1 and even 1 : 0.5 slopes may be achieved by using retaining walls to create earth pockets. Although there are many retaining wall systems available, care should be taken to design a composite structure compatible with the landscape. There are also other earth reinforcing and retaining systems available which incorporate earth as an integral part of the structure. Some of these, which utilise steel mesh and geotextiles, are now being used more frequently and may achieve near vertical profiles (Figure 5.17). Structures that incorporate grass-seed beds, or are hydro-seeded post-construction, can quickly provide a grass wall with appropriate rainfall or watering. Once the grass has become established, shrubs and climbers may be planted to create a more organic and less engineered effect.

The efficacy of the planting to withstand drought needs to be assessed. It is important, however, that an appropriate maintenance programme is agreed and budgeted for in the initial contract so that the planting is watered and weeded until it is well established. Subsequently, during periods of drought, further watering may be necessary. Irrigation should be considered as part of the scheme, especially on steep, retained structures, unless the mound is located in a relatively high rainfall area

**Earth mounds and planting**

Mounds are planted for a number of reasons. The first is to stabilise the structure by binding soil with root growth, so absorbing water running down the slope. The second is aesthetic: to assimilate it into its landscape context. To integrate it better into the landscape, planting should be considered adjacent to and beyond the toe of the mound. Such planting helps

**5.17** Well-maintained grass slopes on a steep reinforced earth mound

Planting on an earth mound can appear unnatural, as planting may emphasise the mound

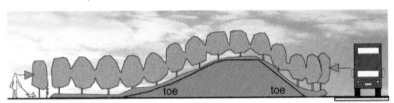

**5.18** Earth mounds and extended planting

Planting extending beyond the toe of the earth mound helps to disguise the mound

to disguise the profile of a potentially imposing form (Figure 5.18). Mound planting also offers an opportunity to create ecological interest and diversity. With the correct aspect and the right soil and slope gradient, mounds can provide sites for wild flowers and other plant species. Ecological importance is increased if insects, moths and butterflies, invertebrates, small mammals and reptiles are also attracted to the site as a result (Figure 5.10).

The type of planting relative to aspect and the angle of a slope is an important consideration for mound design, relevant also to cuttings and embankments. Problems occur where rainfall has decreased and where south- to west-facing slopes (in the UK) dry out in the summer sun so that planting and grass struggle to survive. Plant and grass species need to be chosen very carefully. In some areas, where planting may not survive on a slope, a different type of barrier should be used; planting can then be placed at ground level where the roots can take up ground water. An irrigation system would raise the capital cost but guarantee the long-term success of the design concept.

The second issue, which may affect plant growth, survival, character and quality of planting, is exposure. Factors such as length of growing season, the temperature range and rainfall influence the choice of plant species. Table 5.1 discusses these influences.

The quality of the design and maintenance of planting is an important factor which affects the visual appearance of a transport corridor. It is generally understood that well-maintained and pristine environments can have a beneficial effect, creating a sense of well-being for both travellers and residents. Poor design and lack of maintenance adversely affect the environment and give an appearance of neglect, and ultimately decay. This significantly downgrades the character and quality of the landscape and diminishes the enjoyment of passing travellers. More importantly, it will have a negative effect on the quality of life of local residents. Finally, poor design and maintenance reflect badly on government departments, contractors and design consultants alike.

**Table 5.1** Effects on plants due to angles and aspects[5]

The effects of aspect, especially at steep slope angles, significantly modify local climate in two important respects:

- solar radiation input;
- wind speed and direction, in relation to the local prevailing winds.

These in turn affect the local bioclimate at the ground surface and modify:

- the beginning, duration and end of the growing season;
- the potential evapotranspiration and thus soil moisture balance, particularly the intensity of drought;
- the diurnal temperature fluctuations;
- exposure.

There are few empirical studies of some of the effects, but no sufficiently comprehensive models that can be used to predict their likely extent or intensity with respect to the 'normal' data for a judgement. Some general guidelines are given below.

| Season | Southerly aspects* | Northerly aspects* |
|--------|--------------------|--------------------|
| Winter | Wide range of diurnal temperature variations with regular freeze-thaw cycles. | Narrow range of diurnal temperatures, stays frozen/cold. Snow cover protects vegetation from exposure. |
| Spring | Rapid warming of soil, early start to growing season. Early spells with SMD. | Delayed growing season but it is very rare to experience SMD. |
| Summer | Extreme surface temperatures and very high SMD for extended periods | Moderate surface temperatures, may avoid prolonged SMDs. |
| Autumn | Growing season extends into cooler months. SMD takes longer to be reduced by rainfall. | Early end to growing season, early end to SMD. |

Season prevailing wind conditions should also be taken into account. The angle of south-facing slope receiving maximum solar radiation input:

- Winter               75° from horizontal
- Spring/autumn    55° from horizontal
- Summer             30° from horizontal

* The effect of other aspects will be intermediate between north and south. (The effects noted are for the northern hemisphere.)

SMD = soil moisture deficit.

**Timber barriers**

Timber barriers have been the most frequently used type of barrier alongside roads in the UK. In the past these tended to resemble garden fence structures and have seldom developed a design identity of their own. Although contemporary timber barriers in the UK are generally taller and many of them are absorptive there is little design innovation. A variety of timber barrier types are found in Europe and thus they deserve to be included in the designer's portfolio. They can fit well into the rural landscape and in some cases they can also can appear quite at home in suburban/urban locations; however, they need to be designed beyond the garden fence stage (Figure 5.19). Experience has shown that where barriers are located in residential areas and close to pedestrian routes, they should be on a human scale so as not to appear imposing or threatening. They should appear as light or as natural as possible without creating dark passages and 'scary' places where people are wary of walking (Figure 5.20). In Denmark, for example, timber barriers have been

**5.19** A sympathetically designed timber barrier also serves as a garden boundary

**5.20** An urban timber barrier located within a planter provides a more comfortable space for pedestrians and cyclists

used alongside busy urban roads, but recently these have been rejected by local people in favour of lighter barriers, timber or otherwise, which contain appropriate transparent sections. This is because in some areas where people feel security is an issue, transparent sections make people more visible to one another and thus reduce the chances of attack.

There are, however, locations where timber barriers should be avoided, for example, across viaducts and bridges (Figures 5.21 and 5.22). The use of the material here, outlined against the sky, is inappropriate and ugly. It is not because it is a poor material or acoustically inadequate, but simply its rustic character is out of place on concrete or steel. Its organic nature is better suited to being viewed against a backdrop of planting. Figure 5.22 shows a well-constructed, interesting timber barrier with variation in the profile, texture and colour, but it looks ill conceived when viewed from the approach to the tunnel. In this location there appears to be no real need to screen views of the traffic and a lighter, more sympathetic transparent material could have been used. The designer must reconcile the intrusive nature of the road and traffic with the potentially incongruous nature of the mitigation. In most cases even small timber barriers appear incongruous and aesthetically ill judged in urban situations (5.23 and 5.24). Another situation in which timber should be avoided is where reflective barriers need to be angled for acoustic reasons. Visual expectations dictate that timber fences should be vertical.

Both reflective and absorptive barrier types are available in a wide range of designs. Due to the nature of timber, there is a risk that long stretches of it will be visually boring. Where required for long distances they must be integrated with other barrier types and relieved by planting (Figure 5.25).

**5.21** Timber is a totally inappropriate material for a barrier on this concrete overbridge

**5.22** The incompatibility of timber and concrete is not reconciled by a more complex arrangement of design elements

**5.23** Even small timber barriers should not be located on viaducts. A perforated metal barrier would fit in much better with the concrete structure and metal-clad station canopy

Although timber barriers are ubiquitous in some parts of Europe, they rarely exceed 4–5 m in height, whereas in the UK they are usually 2–3 m tall, but recently barriers have been increasing in height. The continental types appear more varied and robust because of their greater scale and the size of the timber slats used; moreover, they look less like garden fences than British ones since the timber slats are often placed horizontally or diagonally and not vertically. The appearance of these barriers is varied according to the kind of wood that is used, and its colour or staining (Figures 4.25 and 4.91).

**5.24** The small timber barrier has been located to mitigate noise from the train's wheels. Although it maybe acoustically satisfactory, visually it is below par

**5.25** Junction between a timber barrier and a box type bio-barrier

The fact that timber is used for domestic fencing means that people have used timber absorptive barriers in domestic situations. These barriers are found mainly along busy suburban routes. They are simply smaller models of the timber barriers found along motorway and rail routes. They are usually constructed with a solid rear face and fronted with an open façade of battens or overlapping slats containing an absorptive internal section of mineral wool (Figures 5.26–5.29). Most timber barriers are supported by steel I-beams, although in some areas these are substituted by concrete or timber posts.

### Sheet-metal barriers

Sheet-metal barriers are generally absorptive, but reflective ones have also been made. This type of barrier is usually designed using a perforated metal

**5.26** A timber absorptive barrier used as a garden fence in a suburban zone

**5.27** A slatted timber absorptive barrier used in an urban area

**5.28** A louvred timber absorptive barrier used as a garden boundary fence

**5.29** A substantial timber absorptive motorway barrier with concrete posts and decorative lintels

front façade and a solid steel or aluminium rear panel which is not perforated. Aluminium is often chosen in preference to steel as it is lighter and does not rust. The internal space contains mineral wool or other noise-absorbing materials (Figures 5.3 and 5.4). The front façade of most of these barriers is profiled to maximise the strength of the panel and thus increase the span widths between posts (Figures 5.30 and 4.108).

Sheet-metal barriers have been used extensively across Europe, nowhere more so than in Germany, where many have been in place for 25 years or more. The appearance of many of these 4–5 metre tall barriers has stood the test of time, mainly because the materials have weathered well, especially where their appearance is not complicated by inept colour and pattern changes. In many cases, their simple metallic surfaces are graded from dark to light and are softened by planting (Figure 5.31). There are many more elaborate barriers of this type, but their location is better suited to the city, where they are assimilated more readily (Figure 5.32). The use of architectonic features is used to good effect in London on the Docklands Light Railway where structural arches support 2-metre-high vertical and part-cantilevered barriers (Figure 5.33). In future, it may be dismissed as a passing fashion but, currently, it seems that the most visually effective designs are those that respect the integrity of the materials they use. Corten steel, which

**5.30** Detailed view of a profiled sheet-metal barrier with 6 mm diameter holes

**5.31** An older-style perforated sheet-metal absorptive barrier graded in tone and softened by planting

**5.32** Bold sculptural elements used to add character to an otherwise bland cityscape

**5.33** Architectonic features add character to an otherwise plain metal barrier

forms an outer protective rust coating on initial weathering, has been used in contemporary architecture, sculpture and street furniture since the 1980s. It is a material that has effectively been used in an inspired 21st-century barrier at the Craigieburn Bypass outside Melbourne, Australia. The barrier, which utilises a combination of convex and concave Corten steel sheets, terminates in a dramatic bridge structure, sweeping over the carriageways (Figures 5.34–5.36). Keeping the lines simple and strong also helps to create a bold, visually coherent solution (Figures 5.37 and 5.38). The simplicity of form and line may be enlivened, however, by using contrasting materials and colours to accent and punctuate the overall design. In Germany, a 4 metre high barrier on the A2 by Hanover Hospital includes brick entrance features with cast concrete cornicing, bright yellow steel support columns and a decorative top rail (Figures 4.65 and 4.90). These features are continued within an overall design concept which carries through to a bridge across the *autobahn* with a bold yellow façade, and brick and concrete supports (Figure 4.100). Figure 5.39 shows the rear of the barrier and how planting helps to soften its appearance for people enjoying the adjacent woodland.

Sound-absorbent steel and aluminium panels are often combined with transparent panels as well as other sound absorptive systems. Figure 5.40 shows such a system on an Italian motorway outside Milan. The aluminium sound absorptive panels are placed above a sound absorptive New Jersey 'Laghi' barrier, with stepped transparent windows. Sound-absorbent aluminium/steel panels

**5.34** The Craigieburn Bypass Corten steel barrier flows alongside development land as a delightful question mark. Photograph courtesy of Peter Hyatt and Tonkin Zulaikha Greer Architects, Sydney, Australia

**5.35** 'Truth to materials' and delight – A Corten steel barrier enhances the location at the Craigieburn Bypass, Melbourne, Australia. Photograph courtesy of Peter Hyatt and Tonkin Zulaikha Greer Architects, Sydney, Australia

are also suited for placement on retaining walls, retained cuttings and tunnel entrances where they can form part of an overall design concept (Figure 5.41). Apart from a few zigzag barriers (Figures 4.106, 4.107 and 5.75), most metal absorptive barriers are simply profiled or flat. However, an absorptive metal barrier with round sections has been designed for the A2 motorway at Eindhoven.[6] The barrier, which has been conceived to have the potential to also reduce air pollution alongside roads, utilises a series of closely abutted perforated anodised aluminium tubes 3 to 6 m tall (Figures 5.42 and 5.43).

Many of the same design principles apply to reflective sheet-metal barriers. One of the most extensive and visually successful barriers is the 5 metre high aluminium barrier on the A10 Ring around Amsterdam. This imposing

**5.36** The barrier terminates in a dramatic shaped bridge across the bypass. Photograph courtesy of Peter Hyatt and Tonkin Zulaikha Greer Architects, Sydney, Australia

**5.37** Simple lines help to create a bold statement

**5.38** Attention to form and respect for materials creates an appropriate visual character for this sheet-metal barrier

barrier, with a profile that echoes an aircraft wing, is a major architectural structure. Its flowing lines, which follow the curvilinear road layout, give a feeling of dynamism: this high-tech contemporary-looking barrier is a fitting feature on a major route around the city (Figures 5.44 and 5.45). The concept, which hinges on maintaining a seamless, flowing appearance, is further enhanced by narrow, elongated, horizontal windows which are simply used to add interest to a potentially bland, even surface. In total contrast to this, giant climbing frames for planting have been added, which break up the horizontal emphasis and create staccato focal points.

**5.39** Woodland planting visually isolates the barrier for amenity users

**5.40** Aluminium absorptive panels with transparent windows placed on top of a New Jersey 'Laghi' barrier

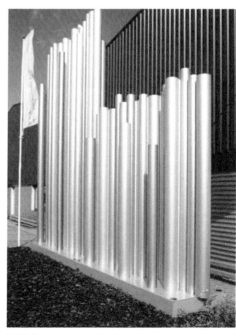

**5.42** Prototype section for a tubular absorptive aluminium barrier destined for Eindhoven, the Netherlands

**5.41** Perforated sheet-metal barrier used at a tunnel entrance

**5.43** Close-up view of the tubular aluminium barrier illustrating perforations which are located on the noise side of the metal tubes

**5.44** A free-flowing functional design in aluminium

**5.45** Climbing frames and horizontal windows add visual complexity to the smooth, even surface

The wing concept again was taken as inspiration for a construction on the A20 in Rotterdam. Here, a strikingly free-flowing design with concealed supports echoes the aesthetic of a dynamic traffic corridor with curvilinear aluminium panels and acrylic windows. The support structure is an elegant wishbone design. Once again, the transport corridor has been seen to present an opportunity for a good functional and aesthetic design that reflects the environment and the age in which we live (Figures 5.46 and 4.24). The unusual use of transparent sections for the lower part of a barrier is successful because it reveals a light and sophisticated support structure.

**Concrete barriers**

Like sheet metal barriers, concrete barriers can be classified according to those that reflect noise and those that absorb noise.

**5.46** A well-conceived and well-implemented architectonic design can improve the motorway setting

**5.47** The visual appearance of concrete can be improved by strong patterning and by softening through planting

### Reflective concrete barriers

Reflective concrete panels and *in situ* concrete constructions can be used as effectively as any other barrier if the overall design is well conceived, the proportions are correct and where planting is used to create an organic, contrasting texture to the concrete. Flat areas of dull concrete should be avoided by using texturing and robustly patterned form-work to create shifting patterns of light and shade (Figure 5.47; see also Figures 4.39, 4.56 and 4.78). Texture and interesting finishes may also be achieved through bush-hammering (Figure 5.48). Massive concrete work without some kind of planting or variation in design should be avoided: compare the massive, visually unaccommodating structure in Figures 5.49 and 4.102 with the examples shown in Figures 5.50 and 5.51. These photographs show how the

**5.48** A bush-hammered profiled concrete surface provides textural interest and changes in visual quality with the play of light and shade

**5.49** A large area of concrete creates an unsightly barrier

**5.50** The visual character of a concrete barrier is relieved by design elements and vegetation

**5.51** Virginia creeper (*Parthenocisus quinquifolia*) used to completely obscure a concrete barrier

**5.52** A concrete-block barrier echoes the vernacular style of the Middle East

appearance of barriers is dramatically improved by an interesting design coupled with planting. The Virginia creeper in Figure 5.51 is planted in an area of soil less than 300 mm wide.

Concrete blocks have also been used to construct noise barriers, an example of which can be found approaching Tel Aviv (Figure 5.52). Here, the colour of the blocks and the pattern tie in with the stone building-block style of construction in Israel; moreover, the rough-textured arrangement of blocks helps to break up a potentially imposing façade.

Most concrete barriers utilise steel I-column posts set within or bolted on concrete foundations. There is, however, one system which requires less extensive footings, by relying on its horizontal alignment to give it structural stability. This system utilises panels which are fanned or snaked and may be constructed up to 10 m or more in height. A greater width is required for the trapezoidal layout, but this may be used for planting (Figures 5.53 and 4.50). These barriers may be reflective or absorptive.

### Absorptive concrete barriers

There are basically two types of absorptive concrete barriers, which may be categorised as woodfibre concrete barriers and granular concrete barriers. Both types usually comprise panels that are made up to size and colour in a factory and are installed between steel supports. The material comprises an open-structured concrete in which either woodfibres or small cementaceous balls are used as the aggregate. The panels are 4–5 m long and usually 140–190 mm thick, depending on whether they are absorptive on one or both sides. Approximately two thirds of the width of the single-sided panel is absorptive whereas the rear third comprises solid concrete. The absorptive surface is usually quite highly profiled in order to increase the surface area of the façade and thus maximise the noise absorption. These profiles can be arranged vertically as well as horizontally or in various patterns using form-work in the factory (Figures 5.54 and 5.55 and 4.39, 4.56 and 5.2). Colour may be incorporated into the panels during manufacture by mixing pigments

**5.53** An 11-metre-high precast concrete articulated 'wave' barrier

**5.54** A profiled woodfibre
concrete barrier

with the concrete mixtures. When viewed from a distance or from a moving vehicle it is difficult to identify which of these two types of absorptive concrete has been used.

Traditionally, woodfibre concrete and granular concrete barriers have been supported by steel or concrete I-columns. Most frequently, the I-columns incorporate a bottom flange which is then bolted onto a steel pile or a concrete footing (Figure 5.56). The panels are placed on concrete footings cast between the column foundations; this ensures that all the panels are perfectly aligned on the same level (Figure 5.57). Joints are normally sealed with compressible rubber strips. The barriers can also be manufactured as complete units which are self-supporting and which can be bolted together to form an apparently seamless barrier. Absorptive and non-absorptive panels are used as part of the modular system being championed in the Netherlands (Figures 5.58 and 4.70).

**5.56** Steel I-column with flange bolted to bolts set into a concrete footing

**5.55** A profiled granular concrete barrier with tree-shaped acrylic windows

**5.57** Precast concrete panels set between steel posts and set onto *in situ* concrete footings

**5.58** Modular woodfibre concrete barrier with 1-metre-tall panels supported at the back with steel support buttresses and elongated footings

### Brick barriers

Bricks are often used to construct masonry barriers as they fit in with vernacular architecture. Solid bricks are used to construct reflective barriers while perforated bricks are used for sound absorptive barriers: either solution creates the impression of a conventional brick wall (Figures 4.65, 4.82, 4.89 and 4.90). These materials can also be used as decorative and functional facings on retaining walls in cuttings. Concrete blocks with perforations can also be used but these present a more engineered and sterile image.

### Plastic, PVC and fibreglass barriers

There are a few examples of barriers that incorporate elements and panels largely composed of plastics, PVC and fibreglass. As plastic recycling increases and these materials become more competitively priced, versatile and robust, it is likely that they will be more widely used. Plastic for many people evokes robust colours and inventive moulded shapes; in this respect, one Parisian barrier exploits the potential of this material with a 5 metre high barrier incorporating a series of flowing, brightly coloured moulded tubes attached to acrylic panels (Figure 5.59). However, plastic barriers need not appear so eccentric for they can be moulded to imitate the character of other materials. For example, the 3 metre high barrier at Nyborg, in southern Denmark, is made from PVC and yet it appears to look like aluminium or coated steel (Figure 5.60; see also Figure 4.77). The barrier is designed with acrylic windows perpendicular to the PVC panels to allow some views out from the motorway.

### Transparent barriers

Transparent noise barriers are constructed from laminated, toughened or reinforced glass or from acrylic or polycarbonate sheet although it is very difficult to differentiate between these materials from a distance. Sheet thicknesses are usually 8–19 mm for glass and 15–20mm for acrylic and polycarbonate. Sheet sizes vary according to the manufacturer, but acrylic sheet can be cut and worked on site whereas glass usually cannot. Very large sheets of acrylic have been used, up to 9 m in height.

**5.59** An older but flamboyant fibreglass barrier

**5.60** A clean-looking PVC barrier, but staggering allows views from the motorway into a private garden (see Figure 4.77)

Acrylic sheet can be curved to add stiffness to the panel, thus maximising span widths between posts and avoiding the use of top rails. This helps to maintain a lightness of form, thereby reducing the apparent height of the barrier (Figure 4.11). Glass barriers, too, can be bent by angling sheets relative to one another to achieve the desired barrier profile. Acrylic sheet can be highly coloured and glass can be subtly tinted and etched (Figures 5.61 and 5.62).

Glass and acrylic sheet are materials which, because of their visual neutrality, have very little impact on the character of the landscape and may be used in most locations where the visual intrusion of traffic is not an overriding issue (Figures, 4.5, 4.6, 4.9, 4.12 and 4.13). Both types of transparent materials are well used across Europe. New planar glazing and curtain walling methods have made it possible to create barriers that are visually high-tech (Figure 4.61). Transparent barriers have also been constructed using glass block units (Figure 5.63).

Two additional factors which influence the choice between glass or acrylic sheet are resistance to vandalism and transparency. Should vandals try to damage a transparent sheet, glass is more easily broken but acrylic sheet is more readily scratched. Where a totally undistorted view through the barrier is required, toughened glass may be preferred to other transparent materials since these may occasionally give rise to minor distortions. Both materials provide transmissibility of light, although historically glass was preferred where light transmissibility over the lifetime of the barrier was important. However, problems of opacity have been overcome with modern acrylics and these now may be guaranteed to maintain their light transmissibility for ten years. A balance must be struck between these factors when selecting the appropriate transparent material.

In most instances, transparent panels have been designed for bridges or viaducts, since their lightweight appearance can be better integrated into the

**5.61** A functional and aesthetic use of acrylic sheet

**5.62** Glass sheets angled to form a cantilevered barrier

**5.63** A unique glass block barrier

engineering design than opaque panels (Figure 5.64). Transparent barriers can usually be built taller and closer to buildings than most other barriers, as they:

- allow access to views, offering no visual obstruction or sense of claustrophobia;
- allow light to penetrate, retaining natural light conditions behind the barrier;
- are generally neutral and visually less intrusive (Figures 5.65 and 5.66 as well as 4.12, 4.13. 4.62–4.64).

Where a barrier is to be placed in an elevated position above a route corridor it makes sense for it to be transparent to maintain views across and beyond the transport corridor (Figure 5.67). This has the added advantage of reducing the visual impact for drivers. Transparent barriers are not, however, always transparent since they are affected by weather and temperature changes. A transparent barrier in the morning may well be semi-opaque until the heat of the day dries off the dew (Figures 5.68 and 5.69).

Where there is close public access to a barrier, it is extremely important to maintain light and a feeling of openness, especially where there is little space available. Figure 5.70 shows a unique 3.5 metre high glazed barrier on the Bern Ostring (East Ring Road) located immediately adjacent to apartment buildings and a public footpath. Adequate noise attenuation is cleverly balanced with the provision of space and light: closely spaced vertical aluminium sound absorptive fins placed at 90° to the front face and a plain, glazed profile to the rear (Figure 5.71).

Another problem for the designer in urban locations is that the design must take account not only of the appearance of the barrier from a distance but also close to: for example, from pedestrian walkways adjacent to the barrier. A barrier may have good lines, shape and form, but to pass close scrutiny it must have sufficient detail and patterning to make it visually interesting. A good example of effective use of shadow to create pattern can be found on a barrier in Copenhagen. As light passes through the silk-screened patterns, the patterns themselves become more intense or diffuse and the

**5.65** A tall acrylic sheet barrier protecting housing and outdoor amenity at Dordrecht, the Netherlands

**5.66** The acrylic sheet barrier is used to create a weather-proof courtyard area for residents while reducing external noise intrusion

shadows cast vary in angle and intensity. The robust, triangular post shape gives a necessary visual gravity to the barrier, but its stocky appearance is broken up by the shadow patterns falling across the posts (Figures 4.105 and 4.22).

The visual character of the supports is important for all barriers, but it can be particularly so with transparent barriers as they will be more noticeable in relation to the glass panels. In many cases, as with other barriers, steel

**5.64** (opposite) An overall confident and well-engineered design includes the barrier within the overall bridge concept

**5.67** A transparent barrier allows views above and beyond the traffic corridor

**5.68** A transparent barrier made non-transparent by morning dew

**5.69** The opaque frost pattern on a transparent barrier ties in with the curved barrier supports creating a serendipitous pleasing effect

I-columns are used (Figure 5.71). This profile adds necessary weight to the visual character of the barrier but also appears crude and heavy. A lightness of appearance can be achieved by tapering the posts, and by angling them to create more space and air at the top of the barrier (Figures 5.72 and 4.18). This lightness of appearance can be further enhanced if no rail is used at the top edge of the transparent panel (Figure 4.11).

Another example where a barrier improves the environmental appearance of a route corridor may be found on the A27 at Gorinchem, Holland. These V-shaped concrete barrier support structures improve the visual quality of the viaduct elevation visible to the neighbouring housing areas, footpaths and cycle-ways. The repetitive use of blue-coloured acrylic panels placed within and above the V creates rhythmic, jewel-like focal points along the viaduct, lifting and strengthening the visual impact of the barrier (Figure 5.73).

**5.70** A transparent barrier maintains light for a pedestrian footpath, whereas perforated aluminium fins provide sound absorption

**5.72** Tapered I-beam tops are more appealing. The detail is carried through into the gantry

**5.71** Viewed close up, I-beams have a heavy appearance

Although coloured glass or acrylic sheet may have advantages, such as being more visible to birds or being less reflective under certain light conditions, their use should be limited to small areas unless used with a reason and with confidence (Figure 4.96). It is much more suitable to locate the colour onto the columns, if colour is required (Figure 5.74).

**5.73** A retrofitted barrier provides noise mitigation and visually enhances an old viaduct

**5.74** The barrier provides a confident architectural statement in the landscape. The columns are coloured and are dominant, whereas the screens are transparent/translucent adding visual detail to an overall bold statement. Photograph courtesy of Peter Hyatt and Tonkin Zulaikha Greer Architects, Sydney, Australia

In most cases, transparent sections are used above other panels to lighten the appearance of the top of the barrier and to reduce its apparent height. In some cases, however, transparent sections may be used as an element in a composite structure whereby the whole appearance of the barrier is lightened. This may be seen in a 3.2 metre high barrier on Route 21, west of Copenhagen, Denmark, where glass and sound absorptive steel panels are sequentially juxtaposed. The steel panels absorb direct sound and sound reflected onto them from the inclined glass panels. At the same time the barrier maintains an open appearance for motorists and allows the properties to the rear of the barrier access to views and light (Figure 5.75).

The extensions either side of the 'Cockpit', which will be discussed in 'Integrated barriers' below comprise a sophisticated plough-shaped skin of glass fixed to a 'TriaGrid'[7] framework. Here, the transparency of the long sections of the barrier has been chosen to enhance the visual qualities of the complex structure and not to allow views to and from the motorway (Figure 4.30).

Transparent barriers and bird strikes

The numbers of birds that are killed by flying into transparent barriers is unknown, but the issue has worldwide relevance as barrier provision is increasing and they are rising in height. This issue is especially crucial in areas where taller barriers may be erected within flight paths and along migration routes. Although the problem is not as severe compared to bird deaths resulting from birds flying into glass buildings,[8][9][10] care should be taken to avoid bird strikes, which mostly end in death. Earlier measures to mitigate bird strikes included the application of large and small black-stencilled raptors onto the transparent sheets of glass or acrylic (Figures 4.97 and 5.6). This practice was superseded by the application of bands of horizontal or vertical striping

**5.75** The clever juxtaposition of materials provides both sound absorptive and transparent qualities

(Figures 4.83, 5.1 and 5.72), but other measures which are used to deter birds from flying into buildings may also be used for barriers. The key to minimising bird strikes is to ensure that the transparent barrier appears solid to birds and that it is apparent that there is no clear passage through the barrier to the open air and habitat beyond. This is mainly achieved through visual markers and by minimising reflections. Measures to prevent bird strike include:

- Bird-strike-resistant glass. Birds are able to perceive ultraviolet (UVA-A) light. The glass utilises a UV-reflective coating which according to the manufacturer is almost invisible to people but visible to birds;[11]
- Create visual markers through patterns – the optimum effective pattern distance is 10 cm or less and clear spaces should not be more than 28 cm.[12] Visual markers can include regular and geometric, vertical and horizontal supports, striping and dots. Visual markers can also include more ornamental and less geometric 'frittering' (Figure 4.23). In acrylic sheet these patterns may be integrated within the sheet's structure with filaments, which also make it shatter-proof when impacted. Patterns on glass are applied, as a film product, silkscreen, and sandblasted or acid-etched;
- Grilles are used on building windows to deter birds. The barrier at Vleuten is designed to deter graffiti, but would also minimise bird strikes (Figure 4.18);
- Photovoltaic cells integrated into glass[13] or as panels can be incorporated into the transparent barriers while also producing renewable energy (Figures 8.2–8.6);
- Reflections can be reduced by angling glass by between 20° and 40°, but this is unlikely to suit the noise mitigation objectives.

## Cantilevered barriers

A cantilevered barrier is one which cantilevers out towards and above the noise source. As discussed elsewhere in the book, they may have a number of visual advantages over and above a simple vertical barrier, as follows:

- reducing overall barrier height by locating the top edge of the barrier closer to the noise source, thereby reducing impact;
- diminishing the impact on the viewer from outside the transport corridor since the top part curves away from the viewer and thus appears lighter (Figures 4.40 and 4.41);
- offering opportunities for bolder, more distinctive design solutions in locations where a boundary fence or wall might appear out of place;
- cantilevering across the carriageway to form a partial tunnel allowing the space above to be used for other purposes.

Cantilevered barriers may be constructed from a range of materials and may vary considerably in height and size. They may be either reflective or sound absorptive. In Denmark, for example, the railway authority uses a comparatively low-key 2.5–3 metre high steel barrier in all locations where noise abatement is required. Although the appearance of the barrier ties in with the utilitarian environment of a railway corridor, it is nevertheless rather dull and uninteresting (Figure 5.76). This utilitarian 'form follows function' design philosophy has also generated the similarly scaled 'Betuweroute' barriers on the freight rail line between Rotterdam, the Netherlands and Germany (Figures 4.75 and 4.102).

At the other end of the scale there are large-scale barriers that are considerable architectural and engineering structures in their own right. The long, 10 metre high concrete cantilevered barrier on the A28 at Zeist, Holland, for example, protects a large estate of high-rise apartments. On the road side, the barrier boldly cantilevers out over the hard shoulder, whereas on the side of the development, robust buttressing counterbalances the ascending cantilevered roof (Figures 2.1 and 2.2). When viewed from the road, the success of this barrier may be attributed to the profile of the roof which arcs smoothly against the sky as it follows the curved alignment of the road. The end profiles are of two different designs: at one end, the façade is angled skywards, giving the profile a dynamic, floating appearance, whereas the other end is staggered in a series of giant steps. Existing woodland has been left untouched between the road and the buildings. The resulting distance between the apartments and the barrier diminishes the impact of such an imposing structure and it is further softened by the retained mature trees. The ordered megalithic concrete buttresses complement the design and echo the pattern of the tree trunks.

Another barrier built on a similar scale was completed on the A16 at Dordrecht, Holland. This 9 metre tall construction screens a complex of older-style apartment blocks and their public spaces from increasing traffic noise levels. The transparent cantilever extends beyond the hard shoulder, supported by a series of prestressed arched concrete columns and a steel lattice frame. The elements comprise dark-coloured profiled concrete panels with intermittent transparent panels at the lower levels. The upper levels are fully transparent. The confident high-tech design, featuring details such as signage structures which curve towards the barrier and echo its shape, signals a resolve to address the increasing noise problems in a fearlessly creative way,

**5.76** A simple small-scale cantilevered railway barrier

entirely in keeping with the urban location (Figures 5.77 and 5.78 as well as Figure 2.4).

Another massive structure which is not, however, as visually appealing, is the 6 metre high sound-absorbent barrier on the A3 at Neudorf, near Duisberg, Germany. This barrier, too, is supported by prestressed concrete columns, but is heavier in appearance, being constructed of granular concrete. Trees behind the barrier help to soften the top profile, but the aspect from the road will inevitably be hard since there is little room for planting; moreover, planting would be difficult to establish and maintain on the front face, or adjacent to it, without irrigation. The choice of the robust Roman copper colour, used successfully elsewhere across Europe, suggests that there has been little attempt to disguise this barrier and initially it appears rather heavy. The problem with the choice of a single colour here is that the cantilever shades the upper portion of the barrier so that the very part that should be lighter in tone appears darker. However, in this situation, a transparent section could not be used because of the need to provide sound absorption (Figure 5.79).

The 6 metre high perforated aluminium absorptive barrier located on the N2 near Bellinzona in southern Switzerland offers a better visual solution to the problem of providing substantial sound-absorbing cantilevered barriers.

**5.77** A well-conceived contemporary design allows a large structure to be acceptable in the landscape

**5.78** Interesting variations in form create a high-tech image

**5.79** The cantilever casts a shadow over the upper elements, where good design principles require lighter tones

Twin cantilevered barriers on either side of the southbound carriageway and a high-tech appearance do not conflict with the semirural, picturesque landscape of the Swiss mountains and lakes. The clean lines of the elegant form wind around the mountain topography to good effect, a model of good design and precise engineering. Although initially surprising, this effect is acceptable in the landscape. Other materials may have been suitable, for example, glass or acrylic, but they would not have provided the required sound absorption. The designers succeeded by applying the principle of using a high-quality design in a high-quality landscape. Planting is used systematically at the front and rear of the barrier, which helps to soften the bottom edges. In fact, the whole design is well conceived, the landscape and road layout outside the barrier forming part of a greater design concept (Figures 4.44, 5.80 and 5.81).

**Thatch barriers**

Thatch seems an unlikely material for a noise barrier, yet panels have been used in the Netherlands. The neutral nature of the material allows them to fit well in the rural landscape, especially where there is a vegetative backdrop. Thatch is fireproofed and placed on a timber panel, supported by steel I-columns and topped with a timber or ceramic cap (Figure 5.82). Frames or wires can be attached to allow climbing plants to create a more organic and natural appearance. Relief patterns can be cut into the thatch as they are on thatched roofs in rural England.

**5.80** A well-designed high-tech solution fits in well even in scenic areas

**5.81** The use of space, planting and the alignment allows the aluminium barrier to fit into the landscape

**5.82** Thatch barriers can fit in well in rural areas

### Bio-barriers

Bio-barriers are structures that incorporate planting as an integral part of their design. Much of the development in bio-barriers occurred in Europe during the 1990s, particularly in the Netherlands. Some early bio-barriers proved unsatisfactory for a number of reasons, such as the need for maintenance and irrigation, but newer designs have addressed these problems. In the UK, however, there is continuing resistance to bio-barriers precisely because of these early teething troubles and the need for continued maintenance. Figures 5.83 and 5.84 illustrate a variety of bio-barrier types and crib and stack systems.

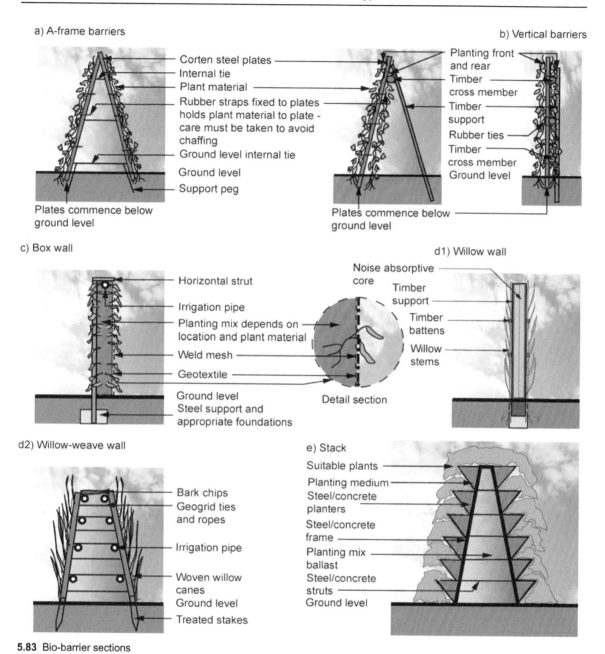

a) A-frame barriers

Corten steel plates
Internal tie
Plant material
Rubber straps fixed to plates
holds plant material to plate -
care must be taken to avoid
chaffing
Ground level internal tie
Ground level
Support peg

Plates commence below
ground level

b) Vertical barriers

Planting front
and rear
Timber
cross member
Timber
support
Rubber ties
Timber
cross member
Ground level

Plates commence below
ground level

c) Box wall

Horizontal strut
Irrigation pipe
Planting mix depends on
location and plant material
Weld mesh
Geotextile
Ground level
Steel support and
appropriate foundations

Detail section

d1) Willow wall

Noise absorptive
core
Timber
support
Timber
battens
Willow
stems

d2) Willow-weave wall

Bark chips
Geogrid ties
and ropes
Irrigation pipe
Woven willow
canes
Ground level
Treated stakes

e) Stack

Suitable plants
Planting medium
Steel/concrete
planters
Steel/concrete
frame
Planting mix
ballast
Steel/concrete
struts
Ground level

5.83  Bio-barrier sections

The question of irrigation and maintenance raises fundamental concerns about attitudes towards barrier provision. Where the decision has been made to provide a barrier to meet legal or design objectives, the appropriate barrier should be chosen and the concomitant maintenance must be regarded as an essential part of the scheme. The need for periodic maintenance should not inhibit this choice. The provision of irrigation enhances plant establishment as well as the continued well-being of plant material over a long period of time.

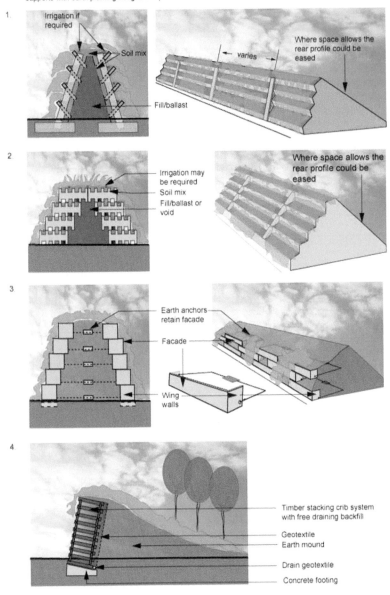

Interlocking systems utilising concrete/timber panels and supports with earth planting troughs and pockets

1.

Irrigation if required

Soil mix

varies

Where space allows the rear profile could be eased

Fill/ballast

2

Irrigation may be required

Soil mix

Fill/ballast or void

Where space allows the rear profile could be eased

3

Earth anchors retain facade

Facade

Wing walls

4

Timber stacking crib system with free draining backfill

Geotextile
Earth mound

Drain geotextile

Concrete footing

Note: In many situations the planting medium will tend to dry out when there is little rainfall or in relation to wind and rain direction. Supplementary irrigation may then be required. Interlocking systems can be double sided or earth mounded on one side.

**5.84** Crib and stacking systems

A range of natural-looking bio-barriers has been developed which offers an alternative to earth mounds. These have the advantage that they do not require the space needed for a mound, in effect creating a living barrier on a narrow strip of land (Figure 5.85). As well as reducing land-take, these bio-barriers act as wildlife corridors creating habitats for small mammals and insects.

Experience has shown that the successful appearance of a bio-barrier depends on:

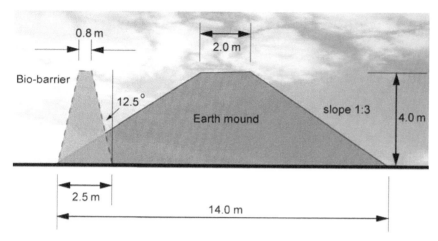

**5.85** Comparative land-take for a 4 metre high earth mound and a 4 metre high bio-barrier

- compatibility of plant species with soil conditions and type (soils must be analysed for fertility, acidity, salinity, contaminants, organic content and drainage);
- density of planting (plants should not compete with one another);
- provision of irrigation or watering during plant establishment;
- provision of irrigation and watering through dry periods;
- establishment of an appropriate plant maintenance regime including weed control, pruning, application of fertilisers and replacement of dead plants.

For ease of categorisation, bio-barriers may be divided into four generic types, the names of which reflect the main structures or principles of the design:

- A-frame and vertical Corten steel bio-barriers;
- box wall bio-barriers;
- woven and non-woven willow bio-barriers;
- stack and crib bio-barriers.

### A-frame and vertical Corten steel bio-barriers

The A-frame barrier consists of two slightly corrugated Corten steel sheets, which are splayed at the base, anchored to the ground with treated timber staves and angled to form an apex. The Corten steel, which forms an anti-corrosive rust on the surface to protect the inner core of the steel, acts as a reflective barrier and is expected to have a useful life of more than 20 years. Plant material is placed immediately adjacent to both sides of the barrier at regular intervals and trained up it using loose rubber ties. Care is taken that the plants do not scrape or chafe against the plates in the wind. Before the planting is established, and during the winter where deciduous planting is used, the steel plates give the barrier a rust colour. This natural colour blends easily into rural/semirural areas. In summer and once the planting is established, the steel is mostly screened and the barrier gives the appearance of a dense, tapering hedge (Figures 5.86 and 5.83a). Thus, deciduous planting

**5.86** A-frame barrier with alder

allows the appearance of the barrier to change colour according to the season in keeping with its surroundings. Planting can be varied according to the location and species such as willow, alder, ash, field maple, privet, lime and ivy have been used in the Netherlands.

The vertical Corten steel barrier comprises a single sheet of corrugated Corten steel which may be placed vertically or splayed slightly. The steel plates are supported by a timber frame to the rear. Planting is usually placed on both sides of the barrier, screening the steel sheets on the front and the frame to the rear. Care should be taken to provide a stable structure since any movement disturbs the plants' roots and inhibits growth (Figure 5.83b).

In the Netherlands and elsewhere in Europe, willow, the most commonly used species with this type of barrier, provides a suitable biotope for some insects. The suitability of willow species, however, should be assessed according to the ground conditions. In areas of low rainfall, irrigation may be needed. Some willows, too, may be sensitive to salt spray and salt in the ground, and susceptible to some diseases and pests, especially in the initial stages of growth. The issue of spray from roads that have been de-iced using salt and the resulting accumulation of salt in the soil is an important consideration for all roadside and barrier planting.

The A-frame and vertical Corten steel bio-barriers appeared in the mid- to late 1990s. These were welcome additions to the designers' portfolio, but as relative newcomers there were a few problems associated with structures, fixings, the long-term establishment of some planting and damage to plants in high winds and other extreme weather conditions. Close planting is not an option since subjects fail due to root competition. Maintenance of the planting and its fixings is an issue complicated by the choice of plants species, the need to train or prune some subjects and to maintain the form of the original design. Although most of these barriers still appear to be functioning and growing well, there has been a recent reluctance to use these barrier types. This is probably due to the maintenance that is required.

Whereas the Corten steel barriers act as reflective barriers, the 'willow wall' barrier (Figure 5.83 d1) utilises willow against a mineral-wool-absorptive core. The core is supported between timber posts and the willow stems are planted into the ground and fixed adjacent to the core with timber battens (Figure 5.87).

### Box wall bio-barriers

The box-type bio-barrier is the most commonly used bio-barrier. It comprises an earth wall within a steel mesh frame with supports. Inside this, the soil mix is contained in a geotextile or polyethylene sheet. The soil mix is adjusted to suit the plant types used. Planting is introduced at intervals in the vertical or slightly tapering sides of the wall through the steel mesh, in holes cut into the geotextile or polyethylene (Figure 5.88). The barrier may be constructed to more than 6 m high, although most existing box-type bio-barriers are 2–4 m

**5.87** A 'willow wall' in early spring

**5.88** Green box-type bio-barrier along a suburban route in the Netherlands

high. Since the barrier is only approximately 0.6 m wide (wider at the base if tapering), it may dry out; irrigation is necessary, therefore. Most barriers in the Netherlands are planted with ivy (*Hedera helix*), which usually grows to completely cover the underlying structure, thus creating an attractive green wall. Their narrow profile and organic but tidy appearance fit into urban, sub-urban and rural areas (Figure 5.83c). They also tie in well with other barriers (Figure 5.25). Soil-less box-type barriers are available which utilise a mineral wool core in which the plants are established. The mineral wool material absorbs moisture from the ground, which passes through the barrier by capillary action; however, although moisture is held within the barrier by capillary breaks, it is advisable to install a dual-purpose irrigation/fertilisation system. In all box systems, planting can be established at the toe of the wall and trained to climb up it (Figure 5.89). If UV-sensitive polyethylene soil containment sheets are used, vigorous root growth is essential to bind the soil as the sheet deteriorates. This problem is avoided if a geotextile is used which is not UV-sensitive and has a longer lifespan.

### Willow-weave bio-barriers

Woven-willow barriers are a patented Dutch invention incorporating vertical staves through which willow whips are woven to form a large cane basket filled with soil, or an appropriate mix. The construction is reinforced using staves, ties and ropes (Figure 5.83d2). Irrigation pipes can be included within the construction. Willow tends to grow quickly, about two metres each season. These whips must be trimmed during the dormant season. Willow walls can create an effective screen but care must be taken that the construction is sufficiently robust. Experience in the Netherlands has demonstrated that many of the horizontal willow whips do not take root and so give the structure no extra stability (Figure 5.90).

**5.89** Newly established box wall bio-barrier, with planting only from the base, which reduces the need for irrigation, but overall plant coverage takes longer. Note 'Devon Banks'-type turf establishment on top

**5.90** A woven-willow barrier awaiting pruning in winter

**5.91** A vertical stacked concrete pipe barrier well integrated into a domestic setting with planting

**5.92** A concrete stack barrier with planting pockets

Stack bio-barriers

The stack barrier is literally a wall of precast units stacked one on top of the other. Most start wider at the base and as the wall steps upwards the number of units in each layer is decreased. In the past, barriers were constructed using pre-cast concrete pipe units, containing ballast and earth in which planting could be established. These developed into a number of different kinds of stackable earth-retaining systems. The overall appearance of these barriers depends on the character and maintenance of the planting (Figures 5.91 and 5.92).

A steel stack system has also been developed. This incorporates a galvanised steel frame on which lightweight earth-filled coated-steel pockets are suspended. These pockets are planted with appropriate plant species. An irrigation system may be required. The base of the unit is approximately 1.6 m wide for a height of 5 m. The barrier faces are angled at 10° from the vertical (Figure 5.93; see also Figure 5.83).

**5.93** A stack barrier using coated steel pockets fixed to a steel frame

Crib wall bio-barriers

Although concrete and timber crib systems are mainly used for earth retention, they can be used as noise barriers by incorporating the cribs on both sides or by using the crib face to steepen the front face of an earth bund. Timber cribs are usually back-filled with a free-draining rock while some pockets are filled with earth and are planted. Most often, however, where planting is required, it is located at the top and encouraged to cascade down. With concrete cribs, soil is usually placed in the pockets to support planting and irrigation is important to help maintain the planting in good order (Figure 5.94; see also Figure 5.84).

**Integrated barriers**

The term integrated barrier is used to identify those barriers which are integrated into the local fabric as utilitarian features. Thus, the first function of the barrier may be a garage, or a storeroom, or a factory unit, while the design and location of the form allows it to act also as a noise barrier. This kind of mitigation is becoming increasingly prevalent in Europe, where new development is proposed alongside busy traffic corridors. In an example at De Lied in the Netherlands, the houses closest to the main road have been provided with a storeroom and open storage area with a large rear wall and sloping roof. These structures are placed in a line at the far end of the gardens facing onto the road. This means that both the gardens and the houses are protected from traffic noise while providing necessary household storage facilities (Figure 5.95). A larger-scale example of this type of barrier may be found in Hong Kong where podium blocks containing car parks and commercial areas are commonly constructed to protect large multistorey housing blocks (Figure 5.96).

**5.94** A fully vegetated concrete crib system provides a large verdant screen

**5.95** A continuous row of storerooms facing the road protects the adjacent homes and gardens

**5.96** Podium block containing car park and shopping areas provides screening for residential tower blocks

The exploitation of the commercial and marketing potentials of integrated barriers is a new phenomenon, which suggests that barrier design is reaching a different kind of maturity. Three projects, two in the Netherlands and the other in Italy, illustrate how development and noise mitigation can be fundamentally interlinked.

The 'Cockpit' barrier in the Leidsche Rijn, on the outskirts of Utrecht, in the Netherlands, has utilised 'high-end' architecture to house a prestigious car dealership, strategically placed along the motorway and close to a motorway junction.[14] The three-storey pod-shaped, glass-covered showroom forms the focus of an elongated, asymmetrical polyhedron, glass-covered barrier on either side of the pod. The 'Cockpit' and barrier structure is constructed out of a 'TriaGrid' system where the façades are used both as load-bearing and aesthetic elements (Figures 4.30, 4.100 and 5.97).

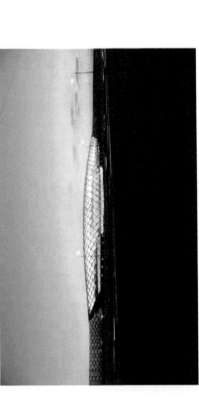

**5.97** The 'Cockpit' incorporates a prestigious car dealership, strategically located alongside the A2 motorway and close to a motorway junction to the west of Utrecht at Maarssen, the Netherlands. The barrier exploits the promotional and logistical advantages of the motorway location while mitigating noise for the development behind it. Cockpit at sunset photograph courtesy of ONL [Oosterhuis_Lénárd], Rotterdam, the Netherlands

'The Wall' also forms part of the greater infrastructure development along the A2 in the Leidsche Rijn district in the Netherlands. The barrier (Figure 4.94), which is under construction at the time of publication, comprises an 800 metre long wall, which will eventually form the eastern façade of a building. The vision of the barrier/building centres on the 'cultural value' of the motorway edge. The building/barrier comprises a concrete stacked form at the base with glass panels above, topped with a large, twisting red facia structure. The whole façade is 13 m tall rising to 25 m tall towards the 'cockpit' barrier where it terminates. The building will contain a shopping mall and at its highest point a restaurant. The clarity and complexity of vision is a visual asset to the motorway edge as it will be in providing noise mitigation and as a cultural and commercial facility for the new residents of the Leidsche Rijn.[15]

Another wall has been constructed along the A4, Milan–Venice motorway near Bergamo (Figure 4.103). The intensely rich, red wall announces the research and office development of Brembo, the makers of vehicle brake sets for luxury vehicles and racing cars. The vibrant red 1,000 metre long wall has an outer skin of extruded and lacquered profiled aluminium, The red wall echoes the corporate image of Brembo and other closely linked Italian red-coloured racing marques. Like the two Dutch barriers discussed above, the design of the Brembo barrier complex grasps the cultural, marketing and corporate potentials of barrier design in addition to the noise reducing facilities that are required.

A further stunning example on the N3 in Dordrecht, the Netherlands, illustrates a fully developed environmental approach to designing a new housing development adjacent to an elevated trunk road. The designers of this large development have used two methods to mitigate noise. The first has been the construction of large 9 m high transparent barriers which

**5.98** A glass and steel barrier provides noise protection as a wind-sheltered outdoor space

enclose the housing area and which, in one part, are attached to the housing to provide sheltered courtyards and conservatories. This is an architecturally dynamic form, that reduces noise and provides a sheltered place to sit, while maintaining views to the lake beyond. Doors in the barrier are provided to allow access to the lake (Figures 5.98–5.100). The environmental design is taken further by locating part of the development in a large man-made bund that faces the road. This large earth structure provides insulation for these houses and noise protection for the rest of the development (Figures 5.100 and 5.101).

The integrated barrier approach can be extended beyond the immediate confines of the noise source. At Lelystad in the Netherlands, a row of houses located some distance away from a busy road has been designed with large reflective roof structures which reflect the noise diffracted over the large

**5.99** Close-up view of the 9 metre high glass barrier that protects the housing behind

earth mound barrier located on the side of the road. The roof structures create an interesting skyline feature and echo the repetitive planting of the avenue of trees (Figure 5.102).

The 'Pontoon Dock' Docklands Light Railway station uses large canti levered structures to not only protect passengers from the weather but also to provide noise mitigation to Thames Barrier Park and existing and proposed development at Silvertown Quays (Figure 5.103, 5.105 and 5.106).

Integrated barriers are also becoming a key element in urban park design.

**5.100** View of the glass-protected housing and earth-sheltered housing to the left. The area in between is protected with a vertical transparent barrier

**5.101** View of the front façades of the earth-sheltered housing illustrated in Figure 5.100

Two projects, one in Paris and one in London, illustrate this phenomenon. At Parc de Bercy (opened 1994/1995), a large retained promenade protects the park from noise, air pollution and visual intrusion from the busy 'Quai de Bercy' dual three-lane highway which is located between the River Seine and the park. The profile on the road side of the park is simple, comprising large classically designed concrete walls. The structure is, however, more varied on the park side. In the southern part, the walls echo the road side but, to the north, the walls give way to an expansive series of steps and a cascading fountain. The top of the noise barrier is designed as a promenade, with views across to the Seine and its west bank. The structure is wide enough to accommodate car parking as well as public toilets (Figure 5.104).

The protection of Thames Barrier Park from noise works in a different way. Here, apart from the central part, the park is raised above the A1020, North Woolwich Road. Noise to the central part, which forms a broad 'cut' within the upper levels, is masked by a series of water jets located between the A1020 and the gardens within the cut (Figures 5.105 and 5.106).

**5.102** Integrated roof/façade design for a row of houses responds to the diffracted noise above a large earth mound, at Lelystad, the Netherlands

### Barrier materials on bridges

Barriers located on viaducts and bridges should, from a visual aesthetic point of view, be kept as light and as simple as possible, but they should also reflect the surrounding landscape character. Where visual intrusion of vehicles is not a significant issue transparent or part-transparent barriers should be used.

The design detail of a noise barrier across a bridge or viaduct should harmonise with the base structure. On bridges, where space is at a premium, noise barriers must be located close to the safety barriers, which can result in visual clutter. The increasing use of concrete parapets as safety barriers avoids this as noise barriers can be mounted on them, achieving a cleaner, simpler appearance (Figure 5.107).

Barrier panels should be tied together and to the structure so that they cannot be easily dislodged if hit by a vehicle or by any other accidental means (Figures 5.108 and 6.1).

Planting on viaducts and bridges is possible and can be used to create an effective visual screen and may also improve the visual character of the structure within its context. Although no examples of this technique are known, noise barriers with integrated planters should be considered, especially if the viaduct extends from one planted area to another (Figures 5.109 and 5.110).

### Barriers and solar panels

At the writing of the first edition of this book, the introduction of solar panels on noise barriers was at an experimental stage and it was suggested that the high costs associated with production of photovoltaics would drop as technology

**5.103** Cantilevered shelters and noise barriers at Pontoon Dock on the Docklands Light Railway

**5.104** Parc de Bercy, Paris: (a) View of part of the park protected by large-scale raised promenade; (b) On top of the promenade in winter; (c) The Quai de Bercy and the protecting wall; (d) Part of the southern park and the park side of the protecting promenade wall; (e) North side of the park with steps up to the promenade and cascading fountain (switched off in winter); (f) One of the access routes to the promenade and access to car parking within the promenade structure

**5.105** View northwards towards North Woolwich Road from Thames Barrier Park, illustrating the higher level of the majority of the park and lower-cut level where noise is masked by water fountains. Note the photograph was recorded before the implementation of the Docklands Light Railway (see Figure 5.24)

**5.106** View southwards into the park towards the Thames Barrier illustrating the higher levels of the park and the cut which is protected from noise by the water fountains illustrated in Figure 5.103

advanced. This has proved true and photovoltaic cells are now being incorporated in increasing numbers of barriers (Figures 4.20 and 5.111). The issue of energy production and noise barriers is discussed more fully in Chapter 8.

## Tunnels

Placing a transport corridor in a tunnel offers the most effective visual and acoustic solution, but it is invariably the most expensive. However, where cut and cover tunnelling is used, the land above can be used for amenity or development purposes, thus defraying the capital cost (Figures 5.112 and

**5.107** Mounting the noise barrier on top of a concrete safety fence avoids visual clutter

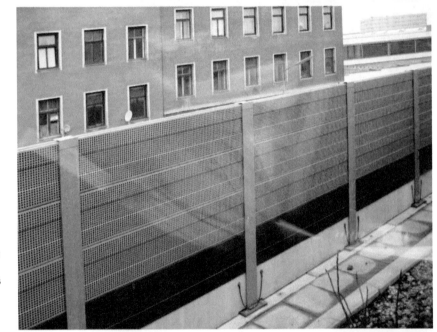

**5.108** Bottom concrete panels of a utilitarian absorptive noise barrier on a railway viaduct linked together for secondary safety in case of accident. Note the top of the I-beam support has been capped, which is visually more appealing than uncapped ends

**5.109** Planting on a bridge structure can link the structure into the surrounding landscape

**5.110** Formal planting on a bridge in an urban environment helps to soften the cityscape

**5.111** Older-style barrier with solar panels placed on top

**5.112** Public open space provided above a motorway hidden below within a cut and cover tunnel

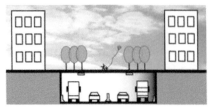

(a) Cut and cover tunnel

Open top with absorptive panels (baffles)

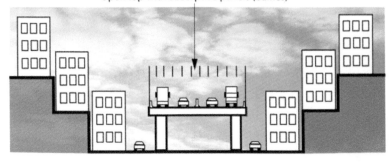

(b) Tunnel on viaduct with absorptive roof baffles

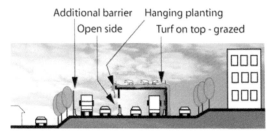

Additional barrier    Hanging planting
Open side          Turf on top - grazed

(c) Open sided gallery above ground level

Previous ground profile dashed line
Additional barrier          Open side    Turf on top - grazed

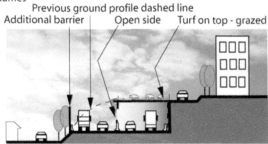

(d) Open sided gallery contained within the landform

**5.113** Tunnels

**5.114** Approaching the elevated Genoese tunnel

**5.115** Inside the Genoese tunnel which contains absorptive roof and side panels

**5.116** The tunnel, which was constructed on an existing viaduct, protects many local residents

5.113a; see also Figure 5.41). The photographs and diagram show the tunnel portal with a perforated aluminium absorptive central reserve barrier at the approach. The covered area has been reclaimed from the traffic corridor to be used as an urban park. Nevertheless, tunnels do not always need to be below ground and nor do they need to be fully enclosed.

A grand example of a tunnel above ground is to be found in the centre of Genoa, Italy (Figure 5.113b). The conditions here are extreme and design options limited: the area is densely developed on rising ground, with multistorey apartment blocks located within meters of a viaduct built by Mussolini in the 1930s. Because of the high volume of traffic the viaduct is now topped with a 270 metre long and 6 metre high, partially open-roofed tunnel. This comprises a louvred cover of vertical sound absorptive panels which allows light into the tunnel, lets exhaust gases escape and attenuates

**5.117** An elevated tunnel at Kings Cross, built for the Channel Tunnel Rail Link into St Pancras railway station in London

**5.118** Galleried tunnel roof tied into the visual character of the surrounding farmland

noise from the traffic. The tunnel sides comprise sound absorptive panels above a concrete New Jersey barrier. The construction is lightweight and is fixed using steel I-columns, steel rods and cables. Autostrade, the organisation responsible for this motorway, reports that local people are happy with the barrier, but that people in apartments level with the barrier have complained about their loss of light (Figures 5.114–5.116).

An elevated tunnel has been constructed to accommodate the Channel Tunnel Rail Link (CTRL), into St Pancras Station in London. Unlike the Genoese tunnel which is rectangular and which echoes the rectangularity of the nearby buildings, the round profile of the elevated CTRL tunnel echoes the profile of the underground tunnels that form much of the link through London and under the Channel (Figure 5.117).

An example of a partial or galleried tunnel is located on the N2 motorway near Lucerne, Switzerland. This long open-sided tunnel, which covers one half of the motorway, is cut into a hillside. At the start the reinforced concrete tunnel stands proud above ground level, but the side walls are well concealed with climbers and shrub and tree planting (Figures 5.113c and 4.92). The roof of the tunnel is turfed over, so blending with the surrounding meadows. The outer edge of the tunnel is planted with shrubs and climbers which hang down over the open section and help to conceal the tunnel from housing on

**5.119** Galleried tunnel fully obscures views of the motorway from houses

**5.120** Unobtrusive escape doors with well-designed signage fits in with the overall barrier concept

the other side. This side is protected by an absorptive perforated metal barrier. Further along, the tunnel is cut into the hillside and the roof of the tunnel is graded into the surrounding land (Figure 5.113d). From there the land rises more significantly and the tunnel roof becomes part of the greater expanse of farmland (Figure 5.118).

The covering of carriageways is becoming more commonplace in Europe. A further urban example may be found near Hamburg airport, where half of a new motorway link road to the airport is covered. The roof area has been turned into an amenity zone and the noise mitigation has been enhanced with an earth mound at the edge of the roof. This mound has been provided with irrigation to help the establishment and maintenance of planting (Figure 5.119).

### Escape routes

Escape doors and routes should be designed as an integral part of the barrier and should tie in with the overall visual concept. It is important, however, that these routes can be easily distinguished and therefore adequate signage and attention-drawing devices should be used without being obtrusive (Figure 5.120; see also Figures 4.39 and 4.78).

### Planting and barriers

(Refer also to Chapter 7.)

Planting is an essential part of environmental design and should be considered in most cases when planning and designing barriers. The planting has a softening, organic effect on an engineered barrier. A planting strategy should be developed with a concept and hierarchy tied into the design of the overall noise and landscape attenuation (Figure 5.121).

Planting can be used in many different ways. It may partially hide the barrier, cover it completely or create focal points. It may be attached to the barrier by climbing wires, frames or ties; located immediately adjacent to it; or be set back from it at various points where it helps to break up the monotony of the façade. Planting can be used innovatively where, for example, post structures are extended into a wire frame which will eventually create a series of vegetated posts (Figure 5.122). Planting can also be located well away from the barrier, possibly off-site by agreement with the landowner, and this helps to diminish its visual impact (Figure 5.123).

Depending on location and landscape character, planting can be organic and natural, or more formally conceived with geometrically designed rows and avenues, or with planters. Barriers can be specially designed to incorporate shrubs and other vegetation (Figure 5.124). Regular planting schemes along the length of the barrier can serve to co-ordinate and harmonise its appearance.

When designing planting immediately adjacent to a barrier care must be taken that the barrier does not cause a rain shadow, restricting the quantity of water reaching the plants' roots. This may be a particular problem with

1.

Climbers - close screening of barrier with self clinging species or climbing wire attached

Maintenance zone may be required on either side

2.

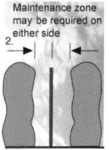

Hedging - formal or informal deciduous or evergreen

3.

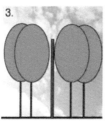

Deciduous trees - partial screening, more screening in summer

4.

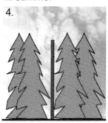

Evergreen trees - most screening usually all year round

5.

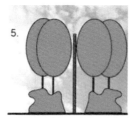

Deciduous trees and shrubs - increased partial screening

6.

Planter with planting - adds height to planting and greater overall vegetation height

7.

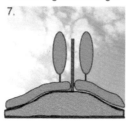

Earth mound adds height to planting and greater screening height

8.

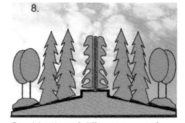

Combination of different types of planting can give good screening

Note: Planting need not be symmetrical; planting may be formal or informal; maintenance requirements may require planting to be located some distance away from the face of the barrier

**5.121** Barriers and vegetation

**5.122** Planting forms part of the support structure

**5.123** Sequential planting helps to break up views to the barrier

larger cantilevered structures that overhang planted areas. In these situations irrigation is essential.

Planting does not necessarily require huge amounts of space. Climbers, both evergreen and deciduous, can be located within widths of 300 mm if care is taken over soil types and drainage requirements. In urban areas, where space is limited, fastigiate trees, or trees of a narrow habit, can be used to advantage (Figures 5.125 and 5.126; see also Figure 5.51).

Well-designed planting is usually a visual asset. It also has the effect of

**5.124** Planting within the barrier planter and adjacent trees help to integrate the barrier into the landscape

**5.125** Tree and shrub planting may be effectively located within a narrow planter

enhancing soil stability, microclimate and the wildlife environment. The mitigation planting strategy and design details should always be agreed with the design team as this may well affect the acoustic mitigation strategy and other environmental issues.

## Barriers and sustainable materials

The issue of sustainability, as well as the carbon footprint and Life Cycle Assessment of products is becoming increasingly important in determining

**5.126** Fastigiate trees used as part of the overall barrier design

**5.127** Detail of an absorptive coir-faced barrier with steel noise isolation plate in the middle

sources and use of material, product and procurement choices. As time passes, the pressure to make sustainable decisions is likely to increase in the built environment including the area of environmental noise barriers. The concept of sustainability is often considered within the framework of 'greenness' or of being 'green'. A number of noise barrier products promote themselves as being 'green'. These so-called green credentials are based mainly on the naturalness of the product, i.e. being made from 'natural' materials such as wood, and being sourced from sustainable/renewable sources. Other barriers pride themselves on being 'green' and natural, being made from materials such as straw (thatch) (see Figure 5.82), and also from waste or as a by-product, such as with the use of coconut husks and fibres. Other green credentials are based on the use of recycled plastics and aluminium.

**5.128** A 4 metre tall barrier. The coir surface provides a good surface for ivy to grow. Photograph two years after planting

A barrier developed in the Netherlands (licensed in the UK) utilises coconut fibres as an offshoot of the coconut/copra industry.[16] The barrier includes a dense metal sheet within or at the rear of panels to reflect noise. In the first instance the road side and, if required (for aesthetic reasons), the community side is packed with noise absorptive, fireproofed coconut fibre (coir) wrapped recycled plastic poles[17] of the type used with pot plants to support small climbers (Figures 5.127 and 5.128). The coconut fibres are sourced from trees growing in salt waters. The fibres are saline and therefore remain clean due to their inherent resistance to algae growth.

This publication does not advocate the use of one material over another as each type may have its rightful niche in a particular situation. However, it is suggested that sustainability and Life Cycle Assessment should inform the choice of material and product.

## References and endnotes

1. The Highways Agency (1992) *Design Manual for Roads and Bridges*, Volume 10, Section 5, Part 2 HA66/95 – Environmental barriers: Technical Requirements, HMSO, London.
2. The Highways Agency (1992) *Design Manual for Roads and Bridges*, Volume 10, Section 5, Part 1 HA65/94 – Design Guide for Environmental Barriers, Section 7, HMSO, London.
3. Roads and Traffic Authority of NSW (1991) *Noise Barriers and Catalogue of Selection Possibilities, Part 1, Noise Barriers*, RTA, Haymarket, NSW.
4. Tutt, P. and Adler, D. (eds) (1985) *New Metric Handbook*, Architectural Press, London.

5. Coppin, N.J. and Richards, I.G. (eds) (1990) *Use of Vegetation in Civil Engineering*, Butterworths and CIRIA (joint pubs.), London, pp. 39–40.

6. Construction is programmed to commence January 2009.

7. The TriaGrid system is a joint venture developed by ONL, Meijers Staalbouw and Centraalstaal in the Netherlands. This system is based on isogrids. An isogrid is a structure which has a sheet of material with integral metal stiffeners in a triangular pattern on one side. Isogrids are fabricated by machining material away from a thick sheet or plate of metal, leaving a continuous flat surface on one side and a series of triangular pockets with thicker integral stiffeners between the pockets on the other.

8. An estimated minimum of one million migratory birds are killed due to glass façade buildings in Toronto, Canada. (City of Toronto Green Development Standard, 'Bird Friendly Development Guidelines', March 2007 – http://www.toronto.ca/lightsout/pdf/development_guidelines.pdf.)

9. An estimated 250,000 birds die each day in Europe by flying into glass buildings.

10. In the USA, it is estimated that migratory and resident bird deaths reach 100 million to one billion birds each year. (U.S.A National Wildlife Federation, 'Steering birds clear of windows' – http://www.nwf.org/nationalwildlife/article.cfm?articleid=52&issueid=9).

11. 'Ornilux' manufactured by Arnoldglas, Merkendorf Germany.

12. City of Toronto Green Development Standard, 'Bird Friendly Development Guidelines', March 2007 – http://www.toronto.ca/lightsout/pdf/ development_guidelines.pdf.

13. The solar cells are embedded between two glass sheets. The void is filled with a special resin wrapping the solar cells on all sides. Each individual cell has two electrical connections which are linked to other cells in the module to form a system that, generates a direct electrical current.

14. The 'Cockpit' barrier and 'The Wall' form part of the strategic infrastructure development for the expansion of housing in the Leidsche Rijn district with more than 30,000 houses and 73,000 people by the year 2015.

15. Ibid.

16. Copra is the dried meat or kernel of the coconut.

17. Recycled mobile phone casings and plastic yoghurt cups.

# Climbing plants and other plants for use on barriers

# 6

## Introduction

This chapter identifies suitable plant species and their integration with environmental noise barriers. Many plant species can be used with barriers. Shrubs and trees may be planted adjacent to barriers to help integration, as well as to facilitate other design objectives such as habitat creation or as architectural focal points. However, the majority of plants that are discussed below are those that can be used as an integral part of barriers. They are the climbers and vines. Yet another group of plants used in barrier design are those that can be planted into the barrier itself. The reasons for using these plants may vary, but the main function of this type of planting is the visual integration of the barrier. In many cases the barrier and its planting are designed to create the impression of a hedge (Figures 6.1 and 6.2; see also Figures 5.51, 5.86 and 5.88).

Wildlife enhancement may also be a secondary, but extremely important, advantage of using planting. For example, a number of bird species such as *Turdus merula* (blackbird in the UK) and *Passer domesticus* (house sparrow in the UK) use climbers such as ivy (*Hedera helix*) for nesting. Bats also sometimes find refuge from the sunlight among its entangled stems, and bees, wasps and flies as well as butterflies feed on the nectar of its flowers. The high fat content of the berries provides a valuable early food source for *Turdus merula* (blackbird), *Columba palumbus* (wood pigeon), *Sturnus vulgaris* (starling) and *Turdus philomelos* (song thrush).

## Considerations for choosing plant species

People often believe that plants reduce noise. While there is evidence to show that leaves can absorb high-frequency sound, this does not significantly affect the propagation of road traffic noise. The perceived reduction in noise

**6.1** Low box-type bio-barrier fully covered in ivy (*Hedera helix*) appears like an evergreen hedge

**6.2** Ivy (*Hedera helix*) creates a dense, glossy green outer surface for the barrier

is largely due to other factors including wind speed and direction, and the sound absorption qualities used for the construction of planted barriers. On the other hand it is important to know that climbers do not generally reduce the acoustic mitigation efficiency of barriers. They do, however, help to trap airborne pollutants and dust.

The characteristics of an appropriate climber for use with environmental noise barriers are:

* medium to fast growth and good longevity;
* ease of establishment;

- disease free;
- largely maintenance free;
- non-fire hazard;
- non-invasiveness;
- available nursery stock and low cost.

Other considerations when choosing climbers:

- there are two types of climbers, those with suckers that cling independently to surfaces and those, that support themselves with tendrils – choose the right type relative to the surface material and provide any appropriate supports such as climbing wires (types and support needs noted for each species below);
- all climbers will need support and training at establishment;
- in some areas using native climbers will enhance ecological values and the use of exotic species will diminish ecological values;
- native species should, if possible, be of local provenance in order to maintain the genetic diversity of the local area. An additional advantage of using locally sourced plants is that they are likely to be hardy in the area;
- all barriers can create 'rain shadows' and plants may not receive sufficient rainfall for growth. Plants must be chosen that can tolerate these local and variable conditions;
- many barriers are located on top of cuttings, mounds and embankments and where rainwater is drawn downwards and away from the plants by gravity. Plants must be chosen that can tolerate these local and variable conditions or irrigation and irrigation maintenance may need to be provided.

There are some locations where it is advisable not to use climbers. These include:

- on transparent and translucent barriers;
- where surface character and quality of the barrier needs to be retained for visual purposes;
- at escape doors;
- where solar panels are being utilised.

## Vertical planting and potentials

Direct planting into barriers has successfully been achieved with the box-type barrier (Figures 6.1 and 6.2; see also Figures 5.83, 5.88 and 5.89). Ivy (mostly *Hedera helix*) is planted directly into the soil contained within the vertical structure. Over time the whole structure is covered and the barrier appears as a hedge. Other species have been successfully used in vertical structures, but these usually require more maintenance and in some cases irrigation. An example of this is the 'green' retaining wall at Thames Barrier Park which uses *Lonicera nitida*, a woody shrub. It is a dense and fast-growing small shrub which can grow in full sun or partial shade (Figures 6.3, 6.4; see also Figure 5.105).

The creation of 'living walls' with vertical planting has become a fashionable architectural accessory in the early 21st century as it creates soft, textural and vital building façades. This type of planting is greatly different in character to the monocultural use of one species, as described above. Natural patterns of light and shade and colour are created by the forms and textural effects of using a variety of ferns, mosses and other foliage plants. The most eminent proponent of these planted walls is Patrick Blanc (Figure 6.5).

The construction of the living walls is similar to the box-type bio-barrier where an inner core contains a planting mixture, contained by a geotextile and typically supported by a reinforcing mesh (Figure 6.5d). In the case of

**6.3** *Lonicera nitida* creates an evergreen retained wall face at Thames Barrier Park

**6.4** Close-up view of *Lonicera nitida* showing construction of the planted wall with reinforcing mesh and geotextile outer structure

**6.5** Living walls: (a) Living wall as an architectural feature at Paradise Park Children's Centre, Islington, London; (b) Living wall by Patrick Blanc. Photograph courtesy of Patrick Blanc; (c) Close up, the wall at Paradise Park Children's Centre is far more textural; (d) Various species can be used but need to have similar microclimate and water requirements – Paradise Park Children's Centre

the living wall on a building this structure is attached to the building. Living walls most often include irrigation. Water is delivered to the top of the wall and then it drips down through the planting mixture where it is collected in a water tray at the bottom of the barrier. The living environmental noise barrier is most often freestanding with internal steel supports (Figure 5.83).

Barriers which use a variety of plant species are likely to only occur in urban situations where they would be viewed at close quarters and where the visual effect can be appreciated (Figure 6.5). However, there are likely numerous ecological advantages of designing living walls in other locations using a variety of plant species. The objective may be to imitate certain habitats and to encourage biodiversity or specific wildlife types. The species used in living walls as architectural features include ferns, mosses, sedges, grasses and numerous types of perennials and small shrubs. (Figure 6.5) The species mix for an environmental noise barrier will need to be designed with specific objectives and the maintenance that will be required for each species.

## Select list of climbers for use in the UK

Table 6.1 (after Grey-Wilson and Matthews[1]) lists deciduous and then evergreen climbers that are suitable for use with barriers. Many other non-native climbing species can be reviewed on the Royal Horticultural Society's plant selector website at: http://www.rhs.org.uk/rhsplantselector/index.aspx.

**Table 6.1** List of climbers

| Latin name of species | Common English name | (H)eight (S)pread in metres | Type of climber: (S)elf-clinging aerial roots or (T)endril/ (Tw)ining | Native to | Sun/Shade ○ Needs full sun ◐ Needs semi-shade ● Needs full shade | Hardiness zone[2] | Comments |
|---|---|---|---|---|---|---|---|
| DECIDUOUS CLIMBERS[3] | | | | | | | |
| *Clematis montana* | Clematis | H 8+ S 3 | Tw Needs support | Himalayas, China | Tolerates ○ ◐ ● | 6–9 | Fast-growing and flowering freely with white to pink flowers |
| *Fallopia baldschuanica* | Russian vine or 'mile-a-minute' | H 6–12 S 4 | Tw Needs support | Asia | ○ or ◐ | 3–8 | Also known as *Polygonum baldscuanica*. Extremely fast and vigorous, rampant twiner with small white and pink flowers. Can appear scraggly in winter |
| *Hydrangea petiolaris* | Climbing hydrangea | H 15 S 3 | S Needs early support | Japan, Korea and Taiwan | Tolerates ○ ◐ ● | 4–9 | Takes time to establish and for aerial roots to provide support. Large white flower caps |
| *Lonicera periclymenum* | Common honeysuckle or woodbine | H 7 S 1.8 | Tw Needs support | Europe, including Britain, from Scandinavia south and east to N. Africa and Greece | ○ ● | 4–10 | Sometime semi-evergreen. A twining scrambling species, best in shade or semi-shaded conditions. Very fragrant – may be a design plus in urban locations |

*Continued*

**Table 6.1** Continued

| Latin name of species | Common English name | (H)eight (S)pread in metres | Type of climber: (S)elf-clinging aerial roots or (T)endril/ (Tw)ining | Native to | Sun/Shade ○ Needs full sun ○ Needs semi-shade ● Needs full shade | Hardiness zone[2] | Comments |
|---|---|---|---|---|---|---|---|
| *Parthenocissus quinquifolia* | Virginia creeper | H 15 S 5 | S But needs some help trained to wall in the beginning | North America | ○ | 3–10 | Vigorous and fast-growing. Striking red foliage in autumn and glossy green in summer. Provides shelter, and occasion nesting space, for the house sparrow, and also for roosting butterflies, such as small tortoise-shells, commas and red admirals. The flowers are valuable to bumble bees for their nectar and pollen[4] |
| *Parthenocissus tricuspidata* | Boston ivy | H 20 S 10 | S Support needed until established | North America | ○ or ● 'Veitchii' ○ | 4–10 | Three-lobed glossy-leaved changing to crimson in autumn. *Veitchii* cultivar tolerates full sun. Provides habitat for insects and birds |
| For *Polygonum baldschuanica* see *Fallopia baldschuanica* above | | | | | | | |

| | | | | | | | |
|---|---|---|---|---|---|---|---|
| *Vitis vinifera* | Common grape vine | H 7<br>S 3 | T<br>Needs support | Asia | ○ or ● | 6–9 | Prefers well-drained loamy soil, slightly calcareous but will succeed on many soils. Useful to blackbird, common wasp, wood mouse, yellow-necked mouse[5] |
| *Vitis cognetiae* | Crimson glory vine | H 10–15<br>S 2 | T<br>Needs support | Japan | ○ or ○ | 5–9 | In autumn turns fiery red, gold and orange. Plants need netting or something similar to cling to. Plants prefer fertile, well-drained, chalky soil |
| *Wisteria floribunda* | Japanese wisteria | H 9<br>S 5 | Tw<br>Needs support | Japan | ○ or ○ | 5–10 | Many hybrids available with purple, pink racemes. Species type is the most hardy. Perhaps best used in urban locations |
| *Wisteria sinensis* | Chinese wisteria | H 30<br>S 5 | Tw<br>Needs support | China | ○ | 5–10 | Very vigorous. Stems can become heavy with age |
| **EVERGREEN CLIMBERS[6]** | | | | | | | |
| *Clematis armandii* | Clematis | H 6–10<br>S 3 | Tw<br>Needs support | China | Tolerates ○ or ○ | 8–10 | Deep glossy green leaves with white/cream/pinkish flowers. Cultivar 'Apple Blossom' is considered by some to be the best with bronzy-green young leaves and pink, blushed white flowers. Very fragrant – may be a design plus in urban locations |

*Continued*

**Table 6.1** Continued

| Latin name of species | Common English name | (H)eight (S)pread in metres | Type of climber: (S)elf-clinging aerial roots or (T)endril/ (Tw)ining | Native to | Sun/Shade ○ Needs full sun ◐ Needs semi-shade ● Needs full shade | Hardiness zone[2] | Comments |
|---|---|---|---|---|---|---|---|
| *Euonymus fortunei* 'Radicans' | Wintercreeper euonymus | H 7 S 1.5 | S With short roots | China | Tolerates ○ ◐ ● | 5–10 | On any soil, pollinated by insects, mature leaves different to immature leaves. Needs initial support until aerial roots establish |
| *Hedera colchica* | Persian ivy | H 6–18 S 5 | S | Iran, Caucasus | Tolerates ○ ◐ ● | 6–10 | Vigorous species with deep green leathery, mainly unlobed leaves. Hybrid 'dentata' has very large unlobed leaves |
| *Hedera helix* | Common ivy or English ivy | H 6–18 S 5 | S | Europe | Tolerates ○ ◐ ● | 5–10 | Many hybrids available. Good for nesting birds and berries eaten by birds. Recommended by the RSPB.[7] Most common species used on environmental noise barriers |
| *Lonicera henryi* | Type of honeysuckle | H 10 S 3 | T | China | ○ or ◐ | | Vigorous non-native. Flowers June/July and black berries provide food for *Parus palustris*, marsh tits and *Parus montanus*, willow tits[8] |
| *Solanum crispum* | Chilean potato vine | H 6 S 4 | Not self-clinging. Needs tying up to wires/trellis for support | Chile | ○ | 8–11 | Usually semi-evergreen. Variety 'Glasvenin' is hardier and very free-flowering with purple flowers and yellow centre |

## References and endnotes

1. Grey-Wilson, Christopher and Matthews, Victoria (1983) *Gardening on Walls*, Collins, London.
2. Tolerates winter temperatures: Zone 1 (–51 to –46 C), Zone 2 (–46 to –40 C), Zone 3 (–40 to –34 C), Zone 4 (–34 to –29 C), Zone 5 (–29 to –23 C), Zone 6 (–23 to –18 C), Zone 7 (–18 to –12 C), Zone 8 (–12 to –7 C), Zone 9 (–7 to –1 C), Zone 10 (–1 to 4 C), Zone 11 (4 to 10 C), Zone 12 (10 to 16 C).
3. Hardiness and whether plants lose their leaves or not during the winter is partially due to the geographic location in which the species will be planted as well as localised conditions.
4. Natural England website – http://www.plantpress.com/wildlife/o986-virginiacreeper.php.
5. Natural England website – http://www.plantpress.com/wildlife/o1019-grapevine.php.
6. Hardiness and whether plants lose their leaves or not during the winter is partially due to the geographic location in which the species will be planted.
7. Royal Society for the Protection of Birds in the UK. '*The holly blue butterflies lay eggs on ivy in the summer . . . Clusters of tiny, five-petalled green flowers appear from late summer through to November, ripening into black leathery berries the following spring. This late flowering season is a valuable source of nectar for many insects prior to hibernation, particularly bees and butterflies . . . the berries of ivy provide many birds, particularly wood pigeon, various thrushes and blackbirds, with abundant food supplies during the most severe months of winter, when little else is available to them.*' Natural England website – http://www.plantpress.com/wildlife/o527-ivy.php.
8. Natural England website – http://www.plantpress.com/wildlife/o1172-marshtitwillowtit.php.

# Engineering, safety, environmental and cost considerations

# 7

## Introduction

To be entirely successful, the design of a noise barrier must address all of the relevant environmental, engineering and safety requirements. The design should be led by the noise control objective and should be visually acceptable while not adversely affecting the landscape character and quality. The other issues must be acknowledged throughout the design, for ultimately statutory and safety considerations will take precedence. Proper provision must also be made in the budget for any barriers that are necessary to meet the design objectives.

The European Committee for Standardisation has prepared a standard covering the performance of the nonacoustic aspects of barrier design and this is published in two parts.[1, 2] The Highways Agency also provides advice on these issues.

## Engineering considerations

The first part of the European standard provides criteria for categorising noise barriers into performance classes. Compliance test methods and reporting procedures are defined for the following aspects:

- wind and static loading, including the effects of dynamic loading due to passing vehicles and static loading due to snow on nonvertical barriers;
- self-weight, including the dry weight to allow an estimate of the sound insulation to be made and, where appropriate, the wet weight;
- impact of stones during normal road use;
- safety of a vehicle in collision with a barrier;
- dynamic load from snow clearance.

General safety and environmental considerations are dealt with in the second part of the standard. The aspects covered are:

- resistance to brush fire;
- secondary safety associated with the risk of falling debris after impact;
- environmental protection with the requirement that any risks posed by barrier materials to the environment over time or at disposal are identified;
- means of escape in emergency, which includes access and egress for personnel and vehicles in emergency and for maintenance;
- light reflection;
- transparency.

Secondary safety is of particular importance where barriers are installed on bridges or between carriageways. Where transparent panels are used, they are made shatterproof by either using laminated glass or by embedding thin strands of fibreglass within acrylic sheets. Panels are tied to each other and to the support posts with short lengths of wire rope at each joint or by using a continuous cable along the length of the barrier (Figures 7.1 and 5.108).

Escape doors and routes are standard features in noise barriers, except for the shortest which allow escape at either end. In the UK the Highways Agency requires doors to be provided at intervals of not more than 200 m, and they should be wide enough to allow stretchers to be carried through. The public should be able to open the doors from the roadside in an emergency, but only maintenance staff and emergency services should have access from the rear. Steps or ramps are provided from the carriageway to the door where it is on a cutting or an embankment. Disabled drivers can be catered for by providing gaps with overlaps in the safety fence and barrier. Vehicle access doors can also be provided for emergency and maintenance vehicles (Figure 7.2). The acoustic performance of the barrier should not be compromised by the presence of escape or access routes (Figures 7.3–7.5; see also Figures 4.39 and 5.120).

The proposed standard does not cover all of the relevant safety and environmental factors and the following should also be considered:

- maintaining the required forward visibility lines for drivers;
- avoiding permanent shadow zones which encourage ice formation;
- sustainability: local materials are likely to produce a more harmonious design solution; moreover, their use should be considered to minimise energy costs associated with imported materials. Similarly local labour should be used wherever possible; the use of materials from non-renewable sources should be avoided.

A further part of the proposed standard will address the long-term durability of barrier materials. The standard will specify test procedures for measuring the resistance to the following agents:

- chemical agents;
- de-icing salts;

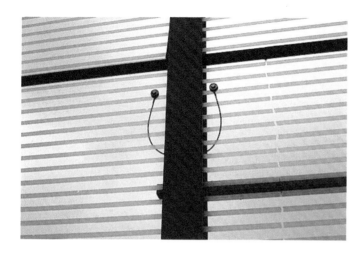

**7.1** Panels are tied to each other and the post for secondary safety

**7.2** Doors for people and vehicles can be combined for escape and access purposes

- dirty water;
- dew;
- freezing and thawing;
- heat;
- UV light.

## Environmental considerations

The construction of an environmental noise barrier may influence a range of environmental issues. These will vary from scheme to scheme and according to the type of barrier used, for example:

- severance – the potential impediment to the existing movement of people or the impediment of existing operations;

**7.3** Overlapping barrier providing secure access route for maintenance on the Betuweroute, the Netherlands

- loss of land – the potential loss of agricultural, amenity and other land-use types;
- archaeology – the potential impact on existing and unknown sites;
- built heritage – the potential impact on listed buildings, National Trust inalienable land, conservation areas, registered parks and gardens and other designated national, county and local sites;
- nature conservation – severance and disruption of habits and habitat of wildlife including mammals, birds and invertebrates, by creating obstacles to paths of travel and effects on designated sites such as SSSIs (Sites of Special Scientific Interest);
- local hydrology – the potential disruption to water bodies and drainage patterns with potential effects on the existing ecology;
- driver stress – the potential impacts on travellers.

Each issue and the potential land-take should be considered separately with the relevant environmental specialist.

The design of a barrier may in fact have some positive effect on the environment and noise barriers may be used to fulfil other environmental objectives. For example, an earth mound may be constructed to fulfil nature conservation objectives to provide habitat for wildlife. Vertical screens can

**Barrier with escape door**

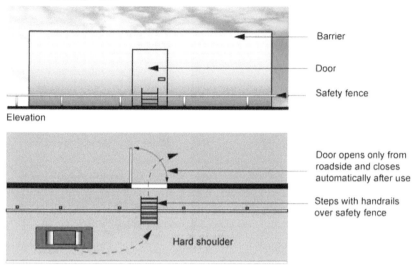

Barrier

Door

Safety fence

Elevation

Door opens only from roadside and closes automatically after use

Steps with handrails over safety fence

Hard shoulder

Plan

**Barrier with escape channel - Option 1**

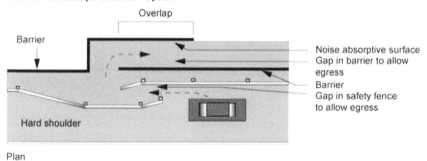

Overlap

Barrier

Noise absorptive surface
Gap in barrier to allow egress
Barrier
Gap in safety fence to allow egress

Hard shoulder

Plan

**Barrier with escape channel - Option 2**

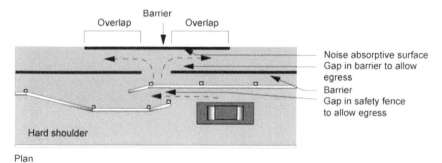

Barrier

Overlap    Overlap

Noise absorptive surface
Gap in barrier to allow egress
Barrier
Gap in safety fence to allow egress

Hard shoulder

Plan

**7.4** Escape doors

be used as part of a nature conservation strategy to stop deer, badgers and other wildlife from straying onto the traffic corridor and to direct them to wildlife bridges or tunnels. (See Chapter 8 where the impact of noise on wildlife and the opportunities are discussed further.)

b)

d)

c)

e)

a)

**7.5** Escape and Access doors: (a) Access door arrangement at a barrier protecting buildings and courtyard. See Figure 8.4. Note the back façade overlaps the opening of the street façade and that double doors allow straight-line access for maintenance vehicles such as a mower; (b) Unlike the previous barrier, there is no fail-safe arrangement to stop noise from passing through the double doors to the apartments and courtyard beyond; (c) The barrier door in this large barrier is at ground level on the road side but somewhat elevated on the protected side. It is assumed that this was the best location for egress; (d) An escape door at the world's largest photovoltaic barrier at Freising, Germany. See Figures 8.1–8.4. Egress is secured but the visual integrity of the barrier is compromised; (e) An escape door within a rock-filled gabion barrier poses construction problems. From an aesthetic point of view, the concrete frame is appropriate but the colour of the door should have been darker and visually recessive

## Graffiti

The motivation of graffiti 'artists' may be unfathomable to most law-abiding citizens, but the fact is that environmental noise barriers often provide an irresistible blank canvas. Graffiti and graffitists rely on two main principles. The first is that the graffiti is visible and the second is that it remains noticeable. Most graffitists will not be bothered to tag or paint where there are no views or where the results will not be visible for a reasonable period of time.

The only effective way of deterring graffiti is through the use of planting. Bio-barriers, which incorporate planting as an integral part of the barrier (discussed in Chapter 5), provide the most difficult target for graffitists and are thus most unlikely to be used for graffiti. Climbers supported on wires or those like ivy (*Hedera helix*), which fix themselves to barriers, screen the barrier surface and provide an organic and difficult façade on which to apply paint. Planting, located in front of barriers, also screens the barriers from view, thus thwarting the artists' main aim, which is exposure. It should be remembered

7.6 Anti-graffiti graffiti: A tram in Amsterdam purposefully painted with its own design used to thwart indiscriminate graffiti

that planting does not require large amounts of space to grow. Ivy (*Hedera helix*) will grow in very small areas of soil in harsh conditions (Figure 5.51). Locating planting within or close to the barrier is not, however, always possible or desirable, and thus for barriers such as transparent barriers on bridges and viaducts, alternative solutions need to be found to deter graffiti.

Design can be used to hinder the graffitist. The barrier at Vleuten, in the Netherlands, provides a contemporary solution alongside an exposed and elevated railway line and station. It uses louvred metal shutters and an inner translucent core (Figures 4.18 and 4.19). Highly profiled barriers also make the graffitists' task that much more difficult (Figures 4.39, 4.56, 5.54 and 5.55).

The main response to graffiti with regard to non-planted barriers or barriers which cannot be screened with planting is to protect the barriers from permanent damage, such as scratching, and to facilitate the removal of paint. A number of services and products have been developed to reduce damage

**7.7** An extensive, uniform barrier readily open to view by many passers-by becomes a target for taggers and graffitists

**7.8** Photomontage of the barrier in Figure 7.7 treated with anti-graffiti graffiti

and to facilitate cleaning. These are of two main types, which relate to the type of surface they are protecting:

1.  Anti-graffiti coatings: These can be applied to most surfaces in the form of a sprayed or painted layer. The mostly water-based coatings inhibit the absorption of the graffiti media on stone, brick, steel and plastics, and thus the paint is more readily removed, usually with high-pressure hot water sprays.
2.  Anti-graffiti films: These are applied to glass and acrylic sheet transparent barriers. The film, which is transparent and thus not easily noticeable by

the graffitist, protects the surface from deliberate as well as accidental damage. The films protect the glass or acrylic sheet below from chemical etching, scratching felt-tip pens and paint. Once the graffitist has despoiled the barrier, the film is easily removed and replaced.

One other method to deter graffitists has been used effectively by 'GVB', the public transport company of Amsterdam. They have applied visually striking designs to the livery of their buses and trams, which limits the visual impact of the graffitists' tag or image (Figure 7.6). This idea could effectively be used to avoid graffiti as the graffitists' main aim is for their work to be clearly seen and identified (Figures 7.7 and 7.8). However, such an approach has its difficulties as the static anti-graffiti design may not fit in with the local context and because graffitists are inventive and may rise to the challenge to overcome the anti-graffiti patterns. No one as yet has identified appropriate anti-graffiti patterns, but they are likely to require a complex arrangement of forms, colours and highlights as well as dark areas.

However, a more positive approach may be beneficial. As part of a marketing and publicity strategy for a new range of footwear, Adidas, in their 'End to End' project, commissioned the painting of a train in Amsterdam by leading graffiti artists (Figure 7.8). Although the graffiti on the train ties in directly with the design motifs on the shoes, it does suggest that in certain cases purposefully designed graffiti may be used to reinforce the identity and visual presence of a commercial enterprise. This idea leads to the potential use of barrier surfaces for advertising. Subtle examples may be found, but these usually take only the form of incorporating a company's name on the barrier (Figures 5.97 and 8.4). In many instances advertising is regarded negatively in townscape, but many people enjoy good-quality advertising. In the appropriate locations these advertising noise barriers could command high rentals, as do many billboards, due to their extraordinary exposure to thousands of vehicle occupants

**7.9** Photomontage illustrating the potentials for advertising on barriers. Compare with Figure 4.50

**Table 7.1** Maintenance cost indicators

| Barrier Type | Factors taken into consideration | Relative cost |
|---|---|---|
| Earth mound | Grass cutting, planting maintenance | Fairly low |
| Timber screen | Inspection/repair, periodic treatment | Low |
| Concrete screen | Inspection/repair, periodic cleaning | Very low |
| Brickwork wall | Inspection/repair, periodic cleaning and repointing | Very low |
| Plastic/planted system | Inspection/repair, periodic cleaning, planting maintenance, irrigation | Moderate |
| Metal panels | Inspection/repair/repainting/ periodic treatment, tighten bolts, check earthing | Fairly low |
| Absorbent panels | Inspection/repair, periodic cleaning | Fairly low |
| Transparent panels | Inspection/repair, regular cleaning/treatment | Fairly high |
| Crib wall | Inspection/repair, planting maintenance | Low |

and viewers (Figure 7.9). The advertising revenues could pay for the barrier over relatively short periods of time. It is suggested, however, that the advertising image(s) would need to change on a regular basis.

## Costs

Cost is probably the main factor which has determined the type and visual character of barriers found in the UK. This, however, does not appear to be the case in Europe, where visual integration and landscape quality considerations are given greater weight. Methods need to be devised to acknowledge the true value of maintaining environmental quality if barriers are to be part of a cost–benefit analysis. If a value is not given to achieving environmental objectives, only the cheapest and inappropriate solutions will be provided.

In most cases barriers have been chosen due to their comparative material costs. The following list of comparative costs and maintenance implications is given by the Highways Agency.[3]

This table attempts to provide information in a simple form. It must be remembered, however, that the location of any barrier will affect the cost of any particular type of barrier. Moreover, the full construction costs can significantly alter the relative barrier costs.

Although it is helpful for budgetary purposes to know what the costs of barrier types are, the budget for each barrier must include all those items necessary to meet the environmental objectives:

- the barrier itself with all its design elements;
- barrier supports and foundations;
- purchase of land for the barrier;
- purchase of land for landscape and other environmental purposes;
- soiling and grading of land;
- planting;
- installation of irrigation.

An adequate maintenance budget must also be provided to ensure the landscape and environmental objectives are met. This may vary depending on the type of barrier required.

Currently, the monetary costs of design and construction are being linked to other environmental costs. Sustainability issues with regard to ecological impacts, materials, energy use in production and transportation and overall carbon footprinting are becoming more important in determining design choices. Cradle-to-cradle costs are one of way of determining which products and methods are best for people and the environment, and it is likely that the positive or negative effects upon the environment, human health and social equity of barrier design and material use will in the future play a great part in barrier choice. Apart from its prime objective, which is to mitigate noise, it is not difficult to envisage that cradle-to-cradle costs are likely to become a dominant determinant in the barrier decision-making process.

## References

1. European Committee for Standardisation (2003) *CEN/TS 1794-1 Road Traffic Noise Reducing Devices. Non-acoustic Performance*. Part 1: Mechanical Performance and Stability Requirements, CEN, Brussels.
2. European Committee for Standardisation (2003) *CEN/TS 1794-2 Road Traffic Noise Reducing Devices. Non-acoustic Performance*. Part 2: General Safety and Environmental Requirements, CEN, Brussels.
3. The Highways Agency (1992) *Design Manual for Roads and Bridges*, Volume 10, Environmental Barriers, Section 5, Part 2 Environmental Barriers: Technical Requirements, HMSO, London, p. 9/3.

# Contemporary issues, developments and considerations  8

## Introduction

When the first edition of this book was published in 1999, environmental noise barrier design appeared to have reached a high level of sophistication, complexity and 'style', particularly in continental Europe. At the time, it became apparent that environmental noise barriers should be considered architecture and not just an addition to road and rail infrastructure or 'street furniture'. Where and when barriers are considered architecture, and furthermore where they are seen to provide commercial and marketing potential, barrier design has progressed with a sense of style and confidence into the 21st century (Figures 4.93, 4.94 and 5.97). It is likely to develop even faster and more dramatically as the century progresses and where commercial and other environmental potentials are realised.

The 21st century has also witnessed accelerated concerns and interest in environmental issues, which were awoken in the latter quarter of the 20th century. The concerns for wildlife, clean air and clean energy are three issues that in the past were not considered part of environmental noise barrier design, but which need to be considered when designing barriers today.

## Noise is an ecological issue

### Introduction

Many animals including mammals (such as badgers and deer), birds (including various fowl, passerines and raptors), amphibians and reptiles (including frogs and newts) are killed on roads in Europe and around the world. Some UK species, such as the great crested newt and the badger, are protected by the Wildlife and Countryside Act 1981. Drivers may often see plastic and wire mesh fencing on the sides of roads which stop these animals from wandering onto the road and being killed. Taller wire mesh and strand fencing is often used to stop deer from crossing roads, but this is in order to protect the impact on drivers rather than the deer themselves, which are not protected. As yet, environmental noise barriers have not been constructed to protect wildlife from noise, but increasing studies have shown noise affects the physical condition and behaviour of animals. Thus, there may be a case for considering the provision of environmental noise barriers to screen important wildlife sites. The US Department of Transportation Federal Highways Administration (FHA), states that 'animals rely on meaningful sounds for communication, navigation, avoiding danger and finding food against a background of noise' and that 'because of the pervasive nature and difficulty in mitigating noise, it may be one of the most significant factor(s) impacting wildlife'.[1]

The FHA notes that the sensitivities of various groups of wildlife can be summarised as:

- Mammals: < 10 Hz to 150 kHz; sensitivity to –20 dB;
- Birds: (more uniform than mammals) 100 Hz to 8–10 kHz; sensitivity at 0–10 dB;
- Reptiles: (poorer than birds) 50 Hz to 2 kHz; sensitivity at 40–50 dB;
- Amphibians: 100 Hz to 2 kHz; sensitivity from 10–60 dB.[2]

With regard to the effects of noise on wildlife, Radle[3] notes that the effects of noise on animals often presents conflicting results due to numerous variables 'such as the characteristics of the noise and its duration, the life history characteristics of the species, habitat type, season, activity at the time of exposure, sex and age of the individual, level of previous exposure, and whether other physical stresses such as drought are occurring around the time of exposure'.[4] Krause's hypothesis states that every animal has its own specific 'aural niche' and sound habitat and that 'the sounds of each of these zones are so unique and important to creature life in a given location . . . that disturbance to this soundscape could be detrimental to the future of the individuals, populations or entire species'.[5] However, she points out that most researchers agree that noise can affect an animal's physiology and behaviour 'and if it becomes a chronic stress, noise can be injurious to an animal's energy budget, reproductive success and long-term survival'.[6]

### Birds

With regard to birds, Forman et al. state that 'traffic noise is interpreted as the overwhelming cause of the underlying correlations of avian patterns with roads and traffic',[7] but 'there is no definitive evidence to explain why noise has a profound effect on some species but not others and at distances that would seem to preclude noise-masking vocalization of up to 3 km'.[8] Research by Brumm[9] notes that male nightingales (Luscinia megarhynchos Brehm) sang louder in noisier locations than in areas with lower background noise levels and that they in fact sang louder on weekday mornings, with increases in traffic compared to weekend mornings.[10] The birds tried to mitigate the deterioration of their communication caused by the intrusive noise.

It is suggested that an increase in vocal strength by the birds may help to maintain a given transmission distance of songs, which are used in the defence of territories and the attraction of mates. However, this comes at a yet unknown cost both physically and in terms of behaviour changes for the male birds. A further example of behaviour modification has been recorded in the city of Sheffield in the UK in 2006. The study showed that the European robin (*Erithacus rubecula*) started to sing at night and that, where these robins sang at night, the noise levels during the day were ten decibels louder than at other sites, equivalent to a doubling in loudness. The birds may also become more stressed if the sound of the city keeps them awake longer. Dr Richard Fuller of Sheffield University furthermore notes that 'if they are singing at night, it is going to take more energy than sleeping, so this may not be good news for them'.[11, 12]

Nesting birds that are noted in studies in the Netherlands to have been affected by noise include lapwing (*Vanellus vanellus*), black-tailed godwit (*Limosa limosa*) and possibly also the redshank (*Haematopus ostralegus*).[13] Studies have also indicated that the number of breeding birds in wooded areas is reduced 'in proportion to the density of traffic on the road'. Reijnen *et al.* report a reduction in the numbers of breeding birds adjacent to a busy highway (30,000–40,000 vehicles/day) at a distance of 300 m[14], and in a three-year study Reijnen and Foppen found a negative effect on 17 of 23 species alongside a road with 40–52,000 cars/day.[15] They do note, however, that this adverse effect was diminished in years in which the overall population size was large. In grasslands, reductions were noted for seven out of 12 passerine species,[16] where the effect is most significant above 50 dB(A) on the verge of the road.[17] The study by Reijnen and Fobben noted that at a traffic volume of 5,000 cars/day most species showed a reduction of 12–56 per cent within 100 m of the road, whereas at distances greater than 100 m only the black-tailed godwit (*Limosa limosa*) and oystercatcher (*Haematopus ostralegus*) showed reduction in density. However, 12 to 52 per cent reductions were found for all species at a traffic volume of 50,000 cars/day at distances of up to 500 m. At distances up to 1,500 m, density reductions of between 14 and 44 per cent were noted for sensitive species, which include both waterfowl (shoveler ducks – *Anas clypeata*) and passerine species (black-tailed godwit – *Limosa limosa*, oystercatcher – *Haematopus ostralegus*, lapwing – *Vanellus vanellus*, skylark – *Alauda arvensis*). In another more extensive study by Reijnen and Fobben, 60 per cent reductions were found in 43 species of woodland birds in both deciduous and coniferous forests.[18] 'Noise was the only factor found to be a significant predictor and the number of cars and distance from the road were significant factors in the number of breeding birds'.[19] The researchers note that there was a reduction in densities of between 20 and 98 per cent up to 250 m away from the road, but that the reasons for the effect are not understood and that it is suggested that the reductions are caused by stress.[20]

It should, however, be remembered that there are some birds, for example crows, that find roadside verges as attractive areas to feed. This is largely due to increased food and sometimes vegetative cover, which may be in total contrast to arable and improved agricultural lands located adjacent to the highways. The highway verges thus provide a habitat lost to agriculture where certain bird species may feed and nest.

### Bird strike

The provision of large and extensive transparent barriers may have an effect on bird populations. The numbers of bird deaths caused by barriers is unknown, but it may be considerable considering the numbers of birds that are killed flying into glass buildings. All barriers and specially transparent barriers should be designed to avoid potential bird strikes. This issue is discussed in more detail under the section on transparent barriers in Chapter 5.

### Other species

Kavaler notes that bats which rely totally on echo-location for locating food are 'unable to find food when interference is produced by natural or mechanical means'.[21] Immel, notes that 'the roar of a dune-buggy engine can temporarily disable a reflexive defense of the desert kangaroo rat against one of its archenemies, the sidewinder rattlesnake. The rat normally can hear the snake at 30 inches, which gives it time to kick sand in the snake's eyes and escape. But the engine noise deafens the rat and virtually eliminates its defensive hearing. Until the rat's normal hearing returns, several days later, the snake often wins in an encounter'.[22]

Studies have indicated that marine life is also affected by noise and not only by the noise from ships and other marine activities but from the land as well. Randle, quoting Quinn,[23] notes that in March 1997 a 40 ft sperm whale became trapped in the inshore waters of the Firth of Forth near Edinburgh, Scotland. Scientists attributed this to traffic noise from the rail and road bridges that traverse the waterway. Although they could not confirm their suspicions, the scientists believed that the clamorous noise made the sperm whale reluctant to return to open waters, which eventually caused it to become stranded in the shallows between the bridges, and resulted in its death.

Although there is little information on the effects of noise on invertebrates the FHA notes that a 'few studies have indicated that several species are sensitive especially to low-frequency vibration'. For example, honeybees 'will stop moving for up to twenty minutes for sounds between 300 Hz and 1 kHz at intensities between 107–120 dB'.[24] Earthworms have been shown to move towards the surface near roadways at low frequencies, exposing them as a food source for birds.[25] Generally, roadsides have been found to provide a habitat for significant numbers of invertebrates including 67 species of insects in the UK.[26] This is probably why birds often use roadsides to feed.

### The potential for environmental noise barriers to enhance biodiversity

Generally, concern for noise effects on fauna centre on the impacts on their natural habitats; however, consideration should also be shown for the effects on wildlife in agricultural as well as more urban locations. It is ecologically and aesthetically unthinkable to imagine our farmland, towns and cities without the sight and sound of animals, and especially birds and their bird-song. Slabbekoorn and Ripmeester suggest that providing environmental noise barriers and the increase in height of existing noise barriers would reduce 'detrimental noise levels' and help birds as well as people.[27] They note that it is 'usually also easier to filter out the bird-relevant frequency components of traffic or industrial noise than to block the lower frequencies'.[28]

It is certain that environmental noise barriers can affect wildlife, in particular the movement of wildlife across roads, but many barriers can also assist in reducing the fatalities of mammals by preventing their easy access onto and across roads. However, this positive attribute is only evident as part of an overall ecological strategy, where mammals such as deer and badgers may be directed along the outer façades of barriers and other protective fencing to suitable crossing points which may include culverts, special tunnels and wildlife bridges. Birds may benefit from environmental barriers by being forced to fly higher above the road surfaces than they would normally and thus being able to better resist the air turbulence caused by fast-moving heavy goods vehicles. However, totally transparent barriers have in the past been responsible for many bird deaths. Today, transparent barriers mostly include integrated or applied striping and other visual deterrents to stop birds from flying into them. The use of the black-silhouetted raptors on transparent barriers is no longer in fashion and has ceased in Switzerland all together (Chapter 5).

The idea of using environmental noise barriers to actively enhance biodiversity has not evolved, yet potentials exist especially with regard to nesting birds and roosting bats. The rear façades of barriers are generally quiet and rarely visited by people. They are areas that are often dry (because of a rain-shadow effect), shaded and well protected. Bird's nest boxes and bat roost boxes could easily be positioned on the rear façades of barriers in appropriate locations, with suitable microclimatic and landscape conditions. The base of barriers at ground level could readily be adapted to include hibernacula for snakes and other hibernating animals, such as toads and frogs. It is essential, however, that any provision for wildlife should take account of any risks associated with wildlife entering the road corridor.

## Environmental noise barriers and energy production

### Photovoltaic barriers

The first environmental noise barrier to incorporate photovoltaic (PV) modules was constructed in Switzerland in 1989 along the A13 motorway. Since then numerous trials have been carried out throughout Europe and around the world (Figure 4.20). Compared to the world's largest 'solar parks', the largest of which is located in Spain, which produce up to 20 MW of power, the PV panels incorporated in environmental noise barriers produce much less.

The largest current PV environmental noise barrier is located alongside 1,200 m of the A92 Autobahn at Freising (outside Munich), Germany (Figures 8.1 and 8.2). The 6,000 m² barrier is located on top of an earth

**8.1** A view of almost the full extent of the largest photovoltaic barrier in the world located at Freising, Germany

**8.2** View from midway along the 1,200 metre long photovoltaic barrier at Freising, Germany

mound. It is managed by the local council[29], and produces approximately 620 kW of power, which is sold back to the grid. The Freising barrier comprises two sections which were built at different stages. The 2002 stage, 320 metre long barrier, comprises 1,000 m² of solar panels producing 122 kW power at a cost of €870,000. The barrier was extended by 890 m in 2003, using 4,000 m² Teflon-coated ceramic solar panels located in five arrays in the top section, as well as two arrays of glass-covered solar panels within

**8.3** Detail showing photovoltaic arrays: Teflon-coated ceramic at the top and glass- covered panels at the bottom

aluminium frames in the bottom section at a cost of €3,750,000 (Figure 8.3). This part of the barrier produces 500 kW. Some problems have occurred with the Teflon-coated ceramic panels. The Teflon is rougher than glass and thus there is more of a problem with dirt sticking to it. Although the barrier works relatively efficiently as an energy source, its effectiveness in reducing noise is diminished by the angling of the barrier of 35° off the horizontal. This angle was derived in response to objections by residents facing the photovoltaic side of the noise barrier who were concerned about traffic noise being reflected by the photovoltaic panels across the motorway in conjunction with noise they were already receiving from Munich Airport.

Energy production with photovoltaic barriers is directly relative to the area covered in PV panels, the latitudinal location of the barrier, local climate and the amount of solar radiation reaching the barrier, the aspect (azimuth) of the barrier, the angle of the barrier relative to the horizontal and the deterioration over time of the barrier, which may become damaged and dirty. The European Commission, Directorate General, Joint Research Centre, PVGIS, PV Estimation Facility website[30] facilitates the geographical assessment of the solar energy resource, which includes the optimum angle for irradiation. The GIS databases encompass the European Subcontinent, the Mediterranean Basin, Africa and South-West Asia.

In Europe, it is envisaged that energy consumption will increase by 20 per cent to the year 2020 and PV use is similarly increasing by 20 per cent. This use is particularly evident in roof top applications, but there are few reasons why photovoltaic applications on barriers should not increase, especially as photovoltaic prices are expected to decrease by 40 per cent by 2010. There is some evidence of the integration of PV into environmental noise barriers but on the whole their use is still exploratory and tentative (Figure 8.4).

**8.4** Photovoltaic cells located at the top of an environmental noise barrier in Vienna provides some energy but also promotes the 'green credentials' of the local energy provider

One major environmental concern with solar technologies is the difficulty in the recycling of heavy metals such as cadmium, which are used in PV cells. There is also an increased use of cadmium sulphide in the production of PV panels, which is replacing the more expensive silicon. A European-wide collection, recycling and recovery system has been established under the 'PV Cycle' scheme,[31] which aims to recover up to 90 per cent of PV waste by 2015. At the moment, there is a voluntary PV take-back scheme, where for example the PV units at Freising will be returned to the manufacturers for recycling and disposal. Should the voluntary scheme be ineffective, it is likely that EU legislation will be established, demanding that manufacturers provide recycling services to customers.

## Environmental noise barriers and air pollution reduction

### Introduction

Another significant negative effect of traffic on busy roads is the introduction of airborne pollutants from vehicle engines. The response to this problem, which affects the physical health of populations living with and travelling through polluted areas, varies across the world (Figures 8.5 and 8.6). In Europe and the UK, air pollution has dramatically improved, but mitigation measures are still required to meet the 1999 EU 'Framework Directive' and so-called later 'Daughter Directives', especially with NOx/NO$_2$ (Nitrous Oxide[32, 33] and Nitrogen Dioxide[34]) and PM$_{10}$.[35] In the UK and continental Europe, legislation controls and influences emissions from engines, but also includes, for example, the taxation of polluting vehicles such as with the

Estimated annual mean background NO2 concentration, 2010 (ugm-3)
Below 5
5 - 10
10 - 15
15 - 20
20 - 30
30 - 40
40 - 50
Above 50
No Data

Mapping from the 'UK Air Quality Archive' with major roads and locations superimposed

**8.5** Map of the estimated annual mean background for nitrogen dioxide, for 2010 in the UK after mapping from the 'UK Air Quality Archive'. Courtesy of AEA Energy and Environment, on behalf of the UK Department for Environment, Food and Rural Affairs and the Devolved Administrations

London Congestion Charge introduced in February 2003. Non-polluting and less polluting vehicles such as electric cars and cars with hybrid engines are exempt from the tax required on entering central and western parts of London. A further robust response for controlling the nitrous oxide and particulates pollution from diesel engine vehicles in Europe has resulted in the establishment of Low Emission Zones (LEZs). A 24-hour, 7-day-a-week LEZ was introduced in February 2008, which deters the most polluting

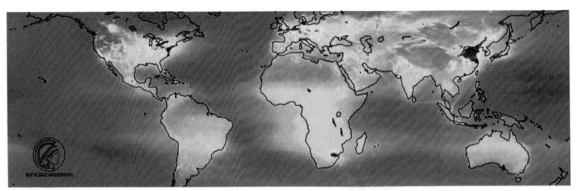

**8.6** World map showing nitrogen dioxide vertical column densities: Mean tropospheric $NO_2$ column density ($10^{15}$ molec/cm$^2$) from measurements of the SCIAMACHY instrument on board the ESA satellite ENVISAT, for the years 2003–2006. Courtesy of Steffen Beirle, MPI Mainz, Germany

diesel engine vehicles (trucks, buses, coaches, minibuses and large vans) from driving within the city of London.

The Netherlands provides numerous exemplars in terms of noise mitigation policy, research and implementation and they are similarly proactive in dealing with traffic generated airborne pollution and by being the progenitors of the CEDR (Conference of European Directors of Roads) air quality group.[36] However, despite the drop in air pollution, it is expected that the European standards for air pollution in the Netherlands will be exceeded and measures are being sought to mitigate this. The costs of meeting the European targets in the Netherlands are estimated to be billions of Euros, through the construction of 'high walls near highways and other screening constructions, tunnels and the removal of houses in the vicinity of highways'.[37] One method for improving air quality that has been proposed is the reduction in traffic speeds.[38] This results from studies on the A13 motorway in the Netherlands with the diminution in air pollution where traffic speeds were reduced to 80 km/h from 100 km/h, with an estimated reduction in 'traffic emissions when compared with the same traffic intensity, with 15–25 per cent for NOx and 25–35 per cent for $PM_{10}$'.[39] Emissions have also decreased by between 4 per cent and 10 per cent on the M42 around Birmingham in the UK where at peak times traffic speeds have been reduced to 50 mp/h (80 km/h) from 70 mp/h (113 km/h) and where the hard shoulder is used as an additional lane during peak traffic flows.[40]

In order to tackle the problem wholeheartedly, the Netherlands government has set up the Innovation Programme Air Quality (IPL).[41] The goal of the IPL programme is to search for and fine-tune cost-effective measures to decrease traffic-related emissions and to help meet the European air quality 'Daughter Directive' in the Netherlands. In this respect, it is noted that 'in contradiction to the topic Noise, there are currently not many known measures to reduce air pollution for roads'.[42] However, measures are being designed to reduce air pollution on and alongside roads and these include the use and modification of environmental noise barriers. The following list includes 11 methods for reducing air pollution alongside roads. The first two utilise environmental noise barriers:

1. Passive Shielding: These measures influence the transfer of traffic emissions to areas surrounding the road. Examples are environmental noise barriers and tunnels. Environmental noise barriers have a diluting effect;
2. Active Dilution: Active decrease with the removal by suction devices, blowing devices or with innovative techniques, such as catalytic roads or catalytic environmental noise barrier surfaces;
3. Faster – Cleaner: Discouraging traffic with relatively high emissions per km and stimulating traffic with relatively low emissions per km. Similar to the Low Emission Zone in London noted above;
4. Volume Limitation: These measures influence traffic intensities at certain locations or on larger traffic networks. Examples of volume limitation are the prohibition of certain vehicle categories on locations and road approach limitation. Similar to the Low Emission Zone in London noted above;
5. Effort on Technology: Promotion of advanced technologies to stimulate development and market of, for example, hybrid vehicles or vehicles with fuel cells;
6. Other Traffic and Transport Sources: Reductions of background concentrations of air pollutants, e.g. railways and shipping;
7. Other Emission Sources: Decreasing emissions by sectors other than traffic and transport, including national and foreign sources;
8. Speed Limit Reduction and Homogenisation of Traffic Flow: Decreasing the dynamics (acceleration and braking) and stimulation of a constant speed of roughly 80 km/hour can expectedly for many road types lead to a decrease of emissions;
9. Environmental Planning and Location of Property: Aims to decrease the exposure of people to high concentrations of pollutants by, for example, creation of zones next to roads without residential allocations;
10. Relocation of Property: Measures in this category consist of relocation of buildings and objects which are close to the road to places further away. An example is the relocation of sensitive receptors like schools and hospitals to locations further from the road;
11. Medical Care: Measures in this category decrease the health risks of air pollution. They do not influence the concentration levels or the amount of exposure. An example would be a programme to increase general public health in order to decrease the health effects of air pollution.

The reduction of air pollution through the use of environmental noise barriers is being tested at Strandnulde in the district of Putten in the Netherlands. Here, a 7 metre tall, 100 metre long barrier and other smaller barriers have been located along the A28 motorway. Air pollution monitors are located 5, 10 and 30 metres behind the barriers (Figure 8.7). A new test barrier with a T-shaped top is also proposed as wind tunnel testing has indicated that it is effective in containing air pollution.

**8.7** Air pollution being tested behind a 7 metre tall test barrier at Strandnulde, the Netherlands

A barrier with a T-shaped top was also submitted as one of the entries in a follow-up competition held in the Netherlands to provoke ideas for environmental noise barriers that also improve air quality and as an extension to the research introduced at Strandnulde (Figure 8.8). The 'AeroStick-T' proposed by the 'Dura Vermeer Groep NV' utilises an air-filtering system at the top of the environmental noise barrier. The 'Una' barrier by 'TNO' utilises half a T-shaped top and an elliptical wing to reduce noise and air pollution through the manipulation of airflows through filters to increase the capture of air pollutants (Figure 8.9 and bottom left Figure 8.8). A number of the proposed barrier prototypes also include catalytic material/chemicals to capture and dissipate the pollutants.[43] The 'Clean Screen' barrier,[44] comprises a T-topped permeable wall, located approximately 60 cm in front of a concrete barrier. The permeable wall contains a gabion-type structure filled with titanium dioxide ($TiO_2$) impregnated lava stones. Electrostatically charged polypropylene filter fins are located between the outer ($TiO_2$) filter and the inner concrete barrier, which provides the noise mitigation element of the environmental noise barrier. Other concept barriers also use panels which are impregnated with $TiO_2$. Titanium dioxide has the ability to accelerate the breakdown of pollutants via a photocatalytic reaction using light as energy.[45] A new and innovative round-section aluminium absorptive barrier destined for Eindhoven has also been conceived for future use in road air pollution mitigation (Figures 5.42 and 5.43).

Practitioners working with environmental noise barriers realise that, in many instances where there is an issue of traffic noise, vehicular traffic is also likely to affect views and the visual quality of an area. In many cases, environmental noise barriers offered solutions to both these issues. The concerns around the world regarding air pollution now indicate that environmental noise barriers have an additional function in the potential for their use in reducing air pollution from roads. It is likely in the following years, as techniques and technologies progress, that the function of reducing air

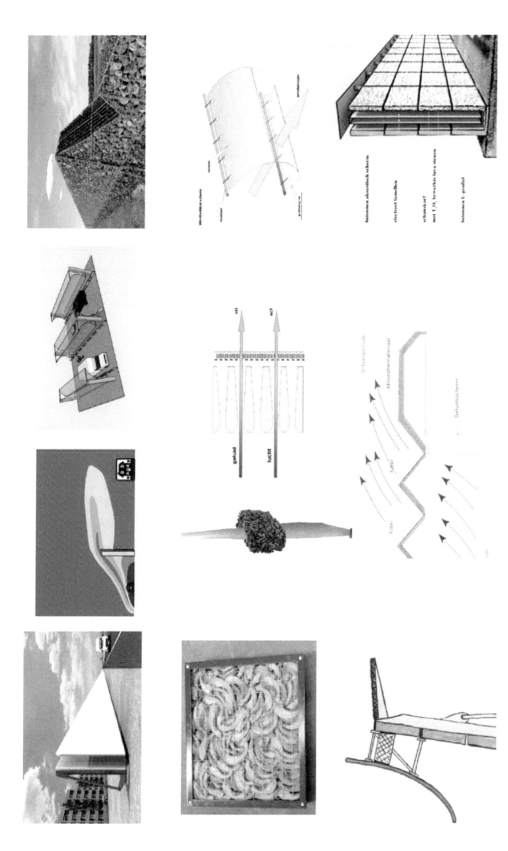

**8.8** A selection of images of environmental noise barriers designed to reduce air pollution submitted into a competition held in the Netherlands in 2008

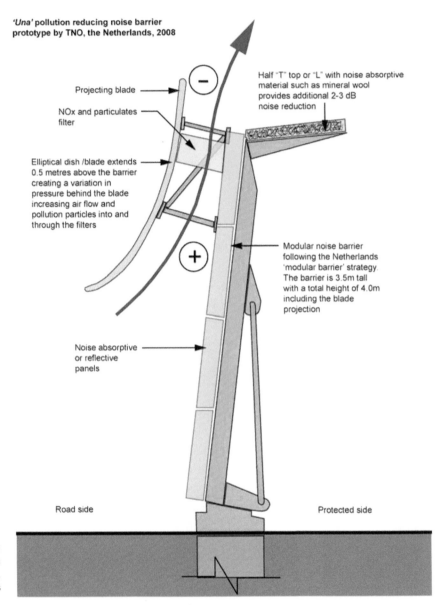

'Una' pollution reducing noise barrier prototype by TNO, the Netherlands, 2008

Projecting blade

NOx and particulates filter

Elliptical dish /blade extends 0.5 metres above the barrier creating a variation in pressure behind the blade increasing air flow and pollution particles into and through the filters

Half "T" top or "L" with noise absorptive material such as mineral wool provides additional 2-3 dB noise reduction

Modular noise barrier following the Netherlands 'modular barrier' strategy. The barrier is 3.5m tall with a total height of 4.0m including the blade projection

Noise absorptive or reflective panels

Road side

Protected side

**8.9** 'Una' prototype noise barrier with air pollution reduction features by TNO, the Netherlands

pollution will become increasingly important. Barrier design will thus have to accommodate three major drivers: noise reduction, the visual screening of traffic and also the reduction of air pollution.

### References and endnotes

1. US Department of Transport, Federal Highways Administration, 'Synthesis of Noise Effects on Wildlife Populations' – http://www.fhwa.dot.gov/environMent/noise/effects/index.htm.
2. Ibid.
3. Autumn Lyn Radle, University of Oregon – http://interact.uoregon.edu/Media Lit/wfae/library/articles/radle_effect_noise_wildlife.pdf.

4. Radle quoting Busnel, R.G. and Fletcher, J. (eds.) (1978) *Effects of Noise on Wildlife*, Academic Press, New York.

5. Autumn Lyn Radle, University of Oregon – http://interact.uoregon.edu/Media Lit/wfae/library/articles/radle_effect_noise_wildlife.pdf

6. Radle quoting Krause – Krause, Bernard. 'The Niche Hypothesis'. The Soundscape Newsletter, 6 June 1993.

7. Forman, R.T.T., Reineking, B. and Hersperger, A.M. (2002) 'Road traffic and nearby grassland bird patterns in a suburbanizing landscape', *Environmental Management* **29**, 782–800.

8. US Department of Transport, Federal Highways Administration, 'Synthesis of Noise Effects on Wildlife Populations' – http://www.fhwa.dot.gov/environ-Ment/noise/effects/index.htm.

9. Brumm, Henrik (2004) 'The impact of environmental noise on song amplitude in a territorial bird', *Journal of Animal Ecology*, 73(3), 434–40.

10. Slabbekoorn, H. and Ripmeester, E.A.P. (2008) 'Birdsong and anthropogenic noise: implications and applications for conservation', *Molecular Ecology* 17, 72–83 quoting Brumm, H. (2004) 'The impact of environmental noise on song amplitude in a territorial bird', *Journal of Animal Ecology*, 73, 434–40.

11. Brahic, C., 'Urban birds sing at night to be heard', *New Scientist Environment*, 25 April 2007 – http://environment.newscientist.com/article/dn11714-urban-birds-sing-at-night-to-be-heard.html.

12. Fuller, R.A., Warren, P.H. and Gaston, K.J., 'Daytime noise predicts nocturnal singing in urban robins', *Biology Letters* (2007/3) pp. 368–70, published online 24 April 2007.

13. van der Zande, A.N., ter Keurs, W.J. and Van der Weijden, W.J. (1980) 'The impact of roads on the densities of four bird species in an open field habitat – evidence of a long-distance effect', *Biological Conservation*, **18**, 299–321.

14. Reijnen, M.J.S.M., Thissen, J.B.M. and Bekker, G.J. (1987) 'Effects of road traffic on woodland breeding bird populations', *Acta Ecologia/Ecologia Generalis*, **8**, 312–13.

15. Reijnen, R. and Foppen, R. (1995), 'The effects of car traffic on breeding bird populations in woodland. IV. Influence of population size on the reduction of density close to the highway', *Journal of Applied Ecology*, **32**, 481–91.

16. Reijnen, R., Foppen, R. and Meeuwsen, H. (1996) 'The effects of car traffic on the density of breeding birds in Dutch Agricultural Grasslands', *Biological Conservation*, **75**, 255–60.

17. Ibid.

18. Reijnen, R., Foppen, R. ter Braak, C. and Thisse, J. (1995) 'The effects of car traffic on breeding bird populations in woodland. III. Reduction in the density in relation to the proximity of main roads', *Journal of Applied Ecology*, **32**, 187–202.

19. Ibid.

20. Ibid.

21. Radle quoting Kavaler, Lucy (1975). *Noise: The New Menace*. New York: The John Day Company.

22. Radle quoting Immel, Richard (1995) 'Shhh . . . Those Peculiar People Are Listening', *Smithsonian*, **26**(1), 151–60.

23. Radle quoting Quinn, Joe 'Whale Trapped in Firth of Forth by Traffic Noise'. Home News, 21 March 1997.

24. Frings, H. and Little, F. (1957) 'Reactions of honey bees in the hive to simple sounds', *Science*, **125**, 122.

25. Tabor, R. (1974) 'Earthworms, crows, vibrations and motorways', *New Scientist*, **62**, 482–3.

26. Free, J.B., Gennard, D. Stevenson, J.H. and Williams, I. (1975) 'Beneficial insects present on a motorway verge', *Biological Conservation*, **8**, 61–72.

27. Slabbekoorn, H. and Ripmeester, E. A. P. (2008) 'Birdsong and anthropogenic noise: implications and applications for conservation', *Molecular Ecology*, **17**, 72–83.

28. Ibid.

29. Freisinger Stadtwerke Versorgungs – GmbH.

30. Refer to http://re.jrc.ec.europa.eu/pvgis/apps/pvest.php?lang=en&map=africa& app=gridconnected.

31. http://www.pvcycle.org/.

32. Nitrogen Dioxide ($NO_2$) – The result of nitric oxide combining with oxygen in the atmosphere, which is a major component of photochemical smog.

33. $NO_2$ is a local air pollution problem. *'Nitrogen oxides (NOx) consist of a mixture of nitrogen dioxide ($NO_2$) and nitrogen monoxide (NO) that are transformed into each other in reaction with other substances.'* One of the main NOx polluters is traffic. People exposed to high levels of nitrogen dioxide are at risk of respiratory ailments, *'to a decrease of respiratory functions . . . increased respiratory complaints and increased reaction to allergens. Epidemiological studies show a positive correlation between weekly $NO_2$-exposure and infection of the airways of children. Long-term exposure of laboratory animals to $NO_2$ shows irreversible changes in the structure and function of the lungs, decrease of the immune system functions and lung emphysema.'* (van Breugel, Peter, 'A search for air quality measures near highways', Road and Hydraulic Engineering Division, the Netherlands – http://ectri.org/YRS05/Papiers/Session-4ter /breugel.pdf.

34. Nitrogen Oxide (NOx) – The result of photochemical reactions of nitric oxide in ambient air; major component of photochemical smog. Product of combustion from transportation and stationary sources and a major contributor to the formation of ozone in the troposphere and to acid deposition.

35. $PM_{10}$: Particulate Matter Up to 10 Microns in Diameter ($PM_{10}$). The number 10 refers to the particle size measured in microns. $PM_{10}$s are widely considered the most dangerous to human health.

36. The group was set up to transfer and co-ordinate knowledge and ensure time and funds were not wasted on similar research in European countries.

37. van Breugel, Peter, 'A search for air quality measures near highways', Road and Hydraulic Engineering Division, the Netherlands – http://ectri.org/YRS05/ Papiers/Session-4ter/breugel.pdf.

38. Reductions in traffic speeds also reduces traffic noise and accidents in most situations.

39. Ibid.

40. Vehicle emission and air quality measurements have shown that vehicle emissions have reduced by between 4 per cent and 10 per cent except for hydrocarbons. Fuel consumption has also reduced by 4 per cent.
    • Carbon Monoxide – reduced by 4 per cent • Oxides of Nitrogen – reduced by 5

per cent • Particulate Matter – reduced by 10 per cent • Carbon Dioxide – reduced by 4 per cent • Fuel consumption – reduced by 4 per cent • Hydrocarbons – 3 per cent increase. (Hydrocarbons do not have an associated Air Quality Strategy Objective and any increase in hydrocarbon emissions will not cause a breach of any legislated limits.) Reductions in speed have also created a minor reduction in noise of between 1.8dB(A) and 2.4 dB(A). (Highways Agency, 'M42 Active Traffic Management Results – First Six Months' – http://www.dft.gov.uk/pgr/roads/tpm/m42activetrafficmanagement/ATM6MonthSummaryResultsforP1.pdf.)

41. Under the Ministry of Transport, Public Works and Water Management and the Ministry of Housing, Spatial Planning and Environment.

42. van Breugel, Peter, 'A search for air quality measures near highways', Road and Hydraulic Engineering Division, the Netherlands – http://ectri.org/YRS05/Papiers/Session-4ter/breugel.pdf.

43. Refer to 'Innovatieprogramma Luchtkwaliteit', Ministerie van Verkeer en Watersaat, the Netherlands – http://www.ipluchtkwaliteit.nl/index.php?page=http%3A//www.ipluchtkwaliteit.nl/project.php%3Fname%3Dschermwerking.

44. Developed by 'van Redubel' in association with 'van Nautilus Schanskorven', 'Keim Nederland B.V.' and '3M Advanced Materials'.

45. Discovered in the 1960s in Japan by Dr Fujishima of Japan. Using energy found in the UV light, photocatalyst titanium dioxide can break down numerous organic and non-organic substances such as oil grime and hydrocarbons from car exhausts It also has hydrophilic properties attracting water from the air. Titanium dioxide is a nontoxic substance found in everyday household products such as toothpaste. Hydroxyl radicals, formed through the chemical reaction of the catalyst and water vapour in the air, destroy the polluting compounds.

# Author Index

# Subject Index

Printed and bound by CPI Group (UK) Ltd, Croydon, CR0 4YY

01/11/2024

01782604-0006